The International Library of Ethics, Law and Technology

Volume 22

Technologies are developing faster and their impact is bigger than ever before. Synergies emerge between formerly independent technologies that trigger accelerated and unpredicted effects. Alongside these technological advances new ethical ideas and powerful moral ideologies have appeared which force us to consider the application of these emerging technologies. In attempting to navigate utopian and dystopian visions of the future, it becomes clear that technological progress and its moral quandaries call for new policies and legislative responses. Against this backdrop, this book series from Springer provides a forum for interdisciplinary discussion and normative analysis of emerging technologies that are likely to have a significant impact on the environment, society and/or humanity. These will include, but be no means limited to nanotechnology, neurotechnology, information technology, biotechnology, weapons and security technology, energy technology, and space-based technologies.

More information about this series at http://www.springer.com/series/7761

Geoffrey S. Holtzman • Elisabeth Hildt
Editors

Does Neuroscience Have Normative Implications?

Springer

Editors
Geoffrey S. Holtzman
New York, NY, USA

Elisabeth Hildt
Center for the Study of Ethics in the Professions
Illinois Institute of Technology
Chicago, IL, USA

ISSN 1875-0044 ISSN 1875-0036 (electronic)
The International Library of Ethics, Law and Technology
ISBN 978-3-030-56133-8 ISBN 978-3-030-56134-5 (eBook)
https://doi.org/10.1007/978-3-030-56134-5

This Springer imprint is published by the registered company Springer Nature Switzerland AG
The registered company address is: Gewerbestrasse 11, 6330 Cham, Switzerland

Introduction

Neuroscience seeks to understand the biological systems that guide human behavior and cognition. Normative ethics, on the other hand, seeks to understand the system of abstract moral principles saying how people ought to behave. Can neuroscience provide insight into normative ethics and help us better understand which human actions and judgments are right and which are wrong? What—if anything—can be learned about normative ethics and philosophical metaethics by studying neuroscientific research? These are the central questions of the collected volume at hand.

In recent years, more and more work in ethics and metaethics has assumed that philosophers can benefit from studying how ordinary people make moral judgments. A growing number of researchers believe that neuroscience can, indeed, provide insights into the questions of philosophical ethics. But exactly what philosophers might be able to learn from such empirical work—and specifically what they can and cannot expect to gain from the neuroscience of moral judgment—is an important but understudied foundational question. Even advocates of the view that neuroscience can provide insights into questions of philosophical ethics acknowledge that the path from neuroscientific *is* to normative *ought* can be quite fraught.

This book consists of a combination of contributions by scholars from disciplines such as philosophy, ethics, neuroscience, psychology, and the social sciences. The collected volume presents different views that all circle around the question of whether neuroscience has normative implications. While some authors in this volume support and embrace the idea that neuroscience does have normative implications and are optimistic about the ways certain neuroscientific insights might advance philosophical ethics, others are more skeptical of the normative significance of neuroscience.

The collected volume begins with chapters that reflect on the role of neuroscience for philosophical thinking on decision-making, moral judgment, moral responsibility, moral cognition, moral motivation, pain, punishment, and social life. In Chap. 1, Jon Leefmann investigates others' claims that some neuroscientific research should be taken as normatively or prescriptively relevant. After distinguishing and discussing an action-theoretic, an epistemological, and a metaphysical reading of this view, he concludes that overall there is relatively limited leeway for inferring concrete normative judgments from neuroscientific evidence.

After identifying and evaluating three perceived threats from neuroscience to our conception of ourselves as free, responsible agents, Myrto Mylopoulos argues that worries for moral responsibility based on these perceived threats are ultimately unfounded. She then suggests ways in which neuroscience, far from serving as a threat, may actually help us to enrich our understanding of ourselves as moral agents.

In Chap. 3, Jennifer Corns and Robert Cowan identify four ethically relevant empirical discoveries about the nature of pain. They then discuss how these discoveries inform putative normative ethical principles and illuminate metaethical debates and conclude that this science-based perspective supports the view that pain is less significant in moral-philosophical contexts than one might have thought.

In "Two Theories of Moral Cognition," Julia Haas uses Fiery Cushman's model-free approach to human cognition as a jumping off point for a novel, multi-system, reinforcement-learning-based model of moral cognition. She argues that moral cognition depends on three or more decision-making systems, with interactions between the systems producing its characteristic sociological, psychological, and phenomenological features.

Chris Zarpentine reflects in his Chap. 5 on the relation between moral judgment and motivation. After examining the dispute between motivation internalists and motivation externalists in light of recent neuroscientific work, he argues that this relation is best seen as a normative one: moral judgment ought to be accompanied by the appropriate motivation.

In Chap. 6, Isaac Wiegman deliberates on normative theories of punishment intended to elucidate why punishment is morally justified. In his Chap. 6, he takes a neuroscience-influenced perspective on retributive and consequentialist considerations regarding punishment and argues that there is less evidence than traditionally thought for the claim that punishment has intrinsic value.

In her Chap. 7, Ullica Segerstrale discusses the potential social consequences of scientific claims about human behavior. Against the background of E.O. Wilson's work, she stresses that neuroscience is vulnerable to the same criticism as sociobiology: the danger of normative interpretations of statements intended to be factual. She particularly criticizes the way the term "tribe" is used and the way in which ingroup-outgroup conflict is presented by some moral psychologists as normal and necessary.

The next three chapters focus on normative implications of neuroscientific findings in medical contexts. Building on KWM Fulford's work, Matthew Ruble argues in Chap. 8 that medical ethics, psychiatric ethics, and neuroethics commit a mistake when attempting to adopt a "facts first then values" approach. He states that a methodology of arguing from allegedly undisputed facts to disputed values is a methodology doomed to moral and epistemic failure.

Christian Ineichen and Markus Christen then discuss the normative implications of neuromodulation technologies used with the aim of pursuing normative goals. After sketching the "standard model" justification of such interventions, which rely on a clear separation between normative considerations and empirical assessments, they challenge this model and provide bridges between the empirical and normative perspective.

In the final Chap. 10, Bongrae Seok explores three different models of interdisciplinary interaction between neuroscience and ethics, focusing on constructive integration. By analyzing recent neuroscientific studies, he argues that neuroscience can be integrated with ethics in developing a normative standard for autistic moral agency.

As a whole, the chapters form a self-reflective body of work that simultaneously seeks to derive normative ethical implications from neuroscience and to question whether and how that may be possible at all. In doing so, the collection brings together psychology, neuroscience, philosophy of mind, ethics, and philosophy of science.

We hope that the volume will be a valuable source not only for philosophers and ethicists interested in philosophy of mind, moral psychology, and neuroethics, but also for psychologists and neuroscientists working on moral cognition. Furthermore, the book could play a central role in graduate courses on neuroethics, moral psychology, philosophy of neuroscience, and philosophy of cognitive science.

Acknowledgments

This collected volumes traces back to the symposium "Does Neuroscience Have Normative Implications?" held in April 2016 at the Illinois Institute of Technology in Chicago, IL. While some of the authors participated in the symposium, we also asked additional scholars to provide their perspectives. The symposium was part of the project "Neuroethics – On the Interplay Between Neuroscience and Ethics." We are grateful to the Cogito Foundation for generously funding the research within the project, the symposium, and the collected volume (Grant 14-108-R). We would also like to thank Kelly Laas for her help with proofreading.

Center for the Study of Ethics in the Professions
Illinois Institute of Technology,
Chicago, IL, USA
Elisabeth Hildt

New York, NY, USA
Geoffrey S. Holtzman
May 2020

Contents

About the Contributors

Markus Christen is Managing Director of the "Digital Society Initiative" of the University of Zurich (UZH) and heads the "Neuro-Ethics-Technology" research group at the Institute for Biomedical Ethics and Medical History at the UZH. His research areas are ethics of information and communication systems, neuroethics, and empirical ethics.

Jennifer Corns is Lecturer in Philosophy at the University of Glasgow. Her published research focuses on pain, affect, and suffering. She aims to use philosophical tools and evaluate empirical research to make progress on topics that matter within and beyond the academy.

Robert Cowan is Lecturer in Philosophy at the University of Glasgow. His research is situated at the intersection of ethics, epistemology, and philosophy of mind. He has recently published papers on topics in these areas in *Canadian Journal of Philosophy, Ethics,* and *Philosophy and Phenomenological Research.*

Julia Haas is an Assistant Professor of Philosophy at Rhodes College. She was previously a McDonnell Postdoctoral Research Fellow in the Philosophy-Neuroscience-Psychology program at Washington University in St. Louis. Her research is in the philosophy of cognitive science and neuroscience.

Elisabeth Hildt is a Professor of Philosophy and Director of the Center for the Study of Ethics in the Professions at Illinois Institute of Technology in Chicago. Her research focus is on bioethics, ethics of technology, and science and technology studies.

Geoffrey S. Holtzman is an independent scholar, who most recently was a Visiting Assistant Professor in the Department of Psychology at Franklin & Marshall College. Previously, Geoffrey was a Postdoctoral Research Fellow at the Geisinger-Bucknell Autism and Developmental Institute and the Geisinger Center for

Translational Bioethics and Healthcare Policy. Prior to that, he was a Postdoctoral Research Fellow in Neuroethics at the Illinois Institute of Technology's Center for the Study of Ethics in the Professions. He received his PhD in Philosophy from CUNY Graduate Center in 2014.

Christian Ineichen is a Postdoctoral Researcher in the Department of Psychiatry, Psychotherapy and Psychosomatics at the Psychiatric Hospital Zurich where he heads in-vivo photometry experiments of depression-relevant behaviors. Besides being affiliated to the Institute for Biomedical Ethics and Medical History at the University of Zurich, he works on behavioral and affective changes after Deep Brain Stimulation in the Department of Neurology at the University Hospital Zurich. His research interests include recording, imaging, and modulation of neural circuits and neuroethics.

Jon Leefmann is Postdoctoral Researcher at the Center for Applied Philosophy of Science and Key Qualifications (ZiWiS), at Friedrich-Alexander-Universität Erlangen-Nürnberg, Germany. His research is mainly concerned with topics at the intersection of applied ethics, epistemology, and philosophy of science. He has published a book and several articles on the ethics of cognitive enhancement as well as several articles on the development of neuroethics as an academic discipline. Currently he pursues a research project on the epistemology of expert testimony and science communication.

Myrto Mylopoulos is an Assistant Professor in the Department of Philosophy, Institute of Cognitive Science, at Carleton University. Her work mainly focuses on issues related to consciousness, agency, skill, and self-control.

Matthew Ruble is a Visiting Assistant Professor of Philosophy in the Department of Philosophy and Religion at Appalachian State University in Boone, North Carolina. He earned both a B.A. in Philosophy and Religion and an M.A. in Psychological Counseling from Appalachian State University. After several years working in community mental health services he returned to school earning an M.A. in Philosophy and Ethics of Mental Health from Warwick University, UK, and a Ph.D. in Philosophy from the University of Tennessee.

Ullica Segerstrale is Professor of Sociology at Illinois Institute of Technology, Chicago. Trained in both science and social science, she is studying such issues as science and social values, scientific conduct and misconduct, and the search for a scientific basis for morality. Among her books are *Defenders of the Truth*, a close-up analysis of the sociobiology controversy (Oxford 2000, 2001), *Beyond the Science Wars* (SUNY Press 2000), and *Nature's Oracle*, a biography of paradigm-changing biologist W. D. (Bill) Hamilton (Oxford 2013, 2015); her work has been supported among others by the Guggenheim and Rockefeller foundations and the American Philosophical Society.

Bongrae Seok is Associate Professor of Philosophy at Alvernia University in Reading Pennsylvania. His primary research interests lie in philosophy of mind, cognitive neuroscience, moral psychology, neuroaesthetics, and Asian comparative philosophy (Confucian moral psychology, Chinese Philosophy, Korean Philosophy). Currently, he is the Chair of Leadership Studies Department and Associate Director of O'Pake Center for Ethics, Leadership and Public Service at Alvernia University. He is the President of ACPA (Association of Chinese Philosophers in America), a member of the Executive Board of NAKPA (North American Korean Philosophy Association), and the Editorial Board of the Korean Society for Cognitive Science.

Isaac Wiegman is a Lecturer in Philosophy at Texas State University who studies the nature and normative significance of evolutionary influences on the human mind. This has led to publications in *Behavioral and Brain Sciences*, *Philosophical Psychology*, *Pacific Philosophical Quarterly*, and *Biological Theory*, among other venues.

Chris Zarpentine received his Ph.D. in Philosophy from Florida State University in 2011. He has taught at the University of Utah and is currently Assistant Professor of Philosophy at Wilkes University. His research adopts an empirically informed perspective and focuses on questions about moral motivation and agency. His work has been published in *Mind*, *Neuroethics*, and *Philosophical Psychology*.

Chapter 1
The Neuroscience of Human Morality: Three Levels of Normative Implications

Jon Leefmann

Abstract Debates about the implications of empirical research in the natural and social sciences for normative disciplines have recently gained new attention. With the widening scope of neuroscientific investigations into human mental activity, decision-making and agency, neuroethicists and neuroscientists have extensively claimed that results from neuroscientific research should be taken as normatively or even prescriptively relevant. In this chapter, I investigate what these claims could possibly amount to. I distinguish and discuss three readings of the thesis that neuroscientific evidence has normative implications: an action-theoretic, an epistemological, and a metaphysical reading. I conclude that the action-theoretic reading has the most direct normative consequences, even though it is limited to the questions of whether some pre-established moral norms can be realized by individual agents. In contrast, in applying the other two readings, neuroscience can only be said to have normative implications in a very indirect way and only under the condition of making contested metaethical assumptions. All in all, the room for inferring concrete normative judgments from neuroscientific evidence is relatively limited.

Keywords Neuroscience of morality · Action · Normativity · Ought-implies-can · Debunking arguments · metaethical naturalism

J. Leefmann (✉)
Zentralinstitut für Wissenschaftsreflexion und Schlüsselqualifikationen (ZiWiS)/Center for Applied Philosophy of Science and Key Qualifications (ZiWiS), Friedrich-Alexander-Universität Erlangen-Nürnberg, Erlangen, Germany
e-mail: jon.leefmann@fau.de

G. S. Holtzman, E. Hildt (eds.), *Does Neuroscience Have Normative Implications?*, The International Library of Ethics, Law and Technology 22,
https://doi.org/10.1007/978-3-030-56134-5_1

1.1 Introduction

In the past 25 years, neuroscientific research as greatly expanded its scope (Matusall et al. 2011; Cacioppo 2016). Among the questions neuroscientists have increasingly addressed are ones about the role of the brain in social and moral behavior (Greene et al. 2001, 2004; Moll et al. 2008; Kahane et al. 2012). Even though there is a significant history of attempts to naturalize human sociality and morality (Spencer 1897; Wilson 1975; Ruse and Richards 2010; Levy 2010; Kitcher 2014), the physiological and the social domains of human life have traditionally been regarded as distinct. There are, however, different levels on which a distinction has been drawn. For an assessment of the prospects of neuroscience to contribute to enlightening and changing social and moral behavior, it is, hence, crucial to distinguish the levels to which claims about the normative relevance of neuroscience apply.

First, one might only want to claim that having agency and being able to control one's behavior depends, among other things, on the workings of one's brain. For a person to have agency implies that she must have the capacity to make decisions, to form intentions and to conduct herself according to them. Hence, the thesis for a role of the brain in social and moral agency can – first of all – be given a reading from the perspective of the theory of action. I will investigate this dimension with regard to the implications of neuroscientific research for our understanding of free agency and moral responsibility.

A second, much more ambitious claim is that neuroscientific evidence can help answer the question of whether certain kinds of moral judgments are justified or unjustified (Greene 2003; Singer 2005). Such a thesis concerns moral epistemology and rests on two important assumptions: First, it assumes that moral judgments can be more or less justified. Second, one can only state of a moral judgment that it is justified or unjustified if there are facts that justify holding the judgment as true. Therefore, one must also make claims about moral ontology. This is to say that if one grants neuroscience to have some implications on the level of recognizing if a moral judgment is justified or not, one also implies that there is some way in which moral facts exist.[1] Some of these implications do not seem to fit well with the ontology typically assumed in the natural sciences, including the neurosciences. According to this ontology there exists nothing over and above those things that can be detected by empirical investigation and explained via the best scientific theories. Hence, if there are moral facts on this view, they can only exist as derivative (or partly constitutive) of the natural (non-moral) facts described by the sciences. From the scientific point of view, moral facts pose a theoretical problem similar to that of

[1] There are several possible views one could adhere to. Besides cognitivism, which will be discussed below, non-cognitivst theories such as quasi-realism might be possible. Also an error-theory of moral judgment according to which humans are only capable of making wrong moral judgments could be an option. Error-theory assumes that there are no moral facts, even though our moral judgments have the property of being either right or wrong (Mackie 1985; Joyce 2007).

mental states, final ends, or social institutions.[2] In all these cases, it needs to be shown how these seemingly non-natural entities can emerge from or supervene on the properties of those entities that make up the furniture of the world of the natural sciences. Of course, solving the metaethical problems of moral realism is by far beyond the scope of neuroscientific investigations of human morality. Yet, the metaethical assumptions implicit to some of such investigations can nevertheless shape the picture one has of human beings and their seemingly distinctive capacity to distinguish right from wrong.

In what follows, I will probe three routes of arguing along this outlined pattern of three different levels of normative implications of neuroscientific research. I do not, however, claim that the sketched routes exhaust all possibilities for reasoning about potential normative implications. There certainly are others. Nevertheless, I will use the outlined approach to argue that neuroscientific research is most likely to affect moral judgments by transforming the use and application of moral concepts. While neuroscience might prove most effective in this regard on the action-theoretic level (Sect. 1.1), its potential to trigger normative reevaluation may also affect attempts to "explain away" certain moral intuitions (Singer 2005). In Sects. 1.2 and 1.3, I will discuss the prospects of such attempts from an epistemological and metaethical point of view. I will argue that the normative implications which in theory could be drawn from neuroscientific investigations into the neural processes accompanying moral judgments can in fact only be made plausible within a broader metaethical framework, which proponents of a neuroscience of human morality should seek to lay open. While it is a philosophical question how such a framework can be made plausible with regard to alternative views in philosophy, I conclude that it will be very unlikely anyway that changes in metaethical theory will affect the way humans should reason in ethics. However, even if this is right, there may be more indirect normative consequences following from the theoretical assumptions and commitments of neuroscientific research into the epistemology and metaethics of human moral behavior. If they leak into extra-scientific, public discourse neuroscientific interpretations of the human capacity to think and act morally – whether plausible or not – can still influence common sense opinion. Which parts of the human experiences with the good or the bad could and should be regarded as morally relevant, may then change in the public eye.

[2]This is not a thesis about the scientific or ontological status of the mentioned entities. The analogy shall only point to the fact that moral facts seem to differ in relevant ways from typical physical entities and that it is at least controversial whether an explanation can be given of them that involves nothing more than the laws of physics.

1.2 Level I: Restricting Moral Agency

From the moral philosopher's viewpoint, the weakest reading of the thesis that the brain has a role to play in explaining human morality is the action-theoretic reading. This reading claims that at least some factors that cause a person to be motivated and capable of following a given norm can be revealed through neuroscientific research. The thesis even seems to be rather uncontroversial. Most people have probably experienced situations of sluggishness, lack of self-control, and lack of self-awareness that were clearly correlated with physiological states such as fatigue, cognitive overload, or hunger. And it would seem rather surprising if a fine-grained neuroscientific analysis of these and similar physiological states would not reveal severe effects on motivation and executive functions.[3] But what do these physiological restrictions of human agency tell about the normative role of neuroscientific evidence?

A fruitful way to think about this question is in terms of moral responsibility. If neuroscience would show that physiological mechanisms that are beyond the agent's control undermine agency in a fashion previously unknown, it seems that our social practice of imputing certain actions to the agent might, in fact, be inappropriate. The least what one could say is that it would be inappropriate to morally blame a person for bringing something about that was beyond her control, and that one cannot properly account to her. What is at issue in this kind of consideration is therefore a kind of retrospective moral responsibility. One can ascribe retrospective moral responsibility to an agent only with respect to actions or consequences of actions that one attributes to that person. This sense of moral responsibility needs to be distinguished from a prospective sense according to which we ascribe moral responsibility to an agent for a certain person, object, or state of affairs by virtue of having certain moral obligations with regard to that person, object, or state of affairs. What is interesting about framing the answer to the question in terms of retrospective moral responsibility is, therefore, that results from neuroscientific research could undermine our practice of holding each other morally responsible for our actions because they question the adequacy of attributing certain actions to the agent. Moral responsibility in the sense at issue, therefore, depends on viewing an agent's behavior as arising from the fact that she has a certain power and capacity to bring about certain states of affairs, and that she has exercised these powers and capacities.

A widely perceived threat to the view that one can attribute an action to the agent, therefore, arises from the observation that humans systematically make mistakes when they do just that. Skepticism arises among others from neuroscientific and psychological experiments.[4] These experiments can be interpreted as showing that

---

[3] Indeed the neuroscience literature on this topic is abundant. See for example: Paschke et al. (2015), Luethi et al. (2016).

[4] The paradigmatic case of this neuroscientific threat to moral responsibility originates from the work of Benjamin Libet and his colleagues. Libet et al. (1983) investigated the timing of brain processes involved in a simple arm movement and compared them to the timing of a consciously

human behavior can be reliably predicted based on certain physiological or psychological features.[5] These features are attributed a causal influence that is independent of the reasons agents refer to in order to explain their behavior. The divergence of the "objective" causes of a behavior and the "subjective" reasons for the behavior are then taken to indicate a lack of awareness of the external stimuli that trigger our actions and, hence, our (self-)ascriptions of moral responsibility. However, the question of how to interpret these findings with regard to an agent's retrospective responsibility has been debated to the point where everything has been said. It is, therefore, hardly surprising that the experiments have been cited in support of (or at least as not contradicting) very different metaphysical views in the debate of free will.

An interpretation of these experiments that I will label the folk-interpretation is to see their results as undermining moral responsibility in virtue of undermining the libertarian conception of free will (Haggard and Eimer 1999). That our behavior is actually determined by a predictable physiological or psychological mechanism is then interpreted as indicating the truth of determinism. This interpretation, however, misses the difference between a process's predictability and a process being determined. That one can predict the outcome of certain brain processes does not indicate anything about determinism as a metaphysical thesis. Even in an overall indeterminate world, there might be predictable processes. Moreover, the fact that we are often mistaken about the physical causes of our behavior is not so much an argument for determinism as it is an argument for the fallibility of our beliefs. One can be wrong about the causes of one's behavior independent of whether one is able to do otherwise.

---

experienced will in relation to self-initiated voluntary acts. They found that the conscious intention to move the arm came 200 milliseconds before the motor act, but 350–400 milliseconds after readiness potential —a buildup of electrical activity that occurs in the brain and precedes actual movement. From this Libet and others concluded that the conscious decision to move the arm cannot be the true cause of the arm's movement, because it comes too late in the neuropsychological sequence and that we are systematically mistaken if we attribute to ourselves that status of being the originator of our arm's movement. From there, it is only one tiny step also to draw the conclusion that we would not be morally responsible in the retrospective sense should the arm's movement cause morally problematic consequences. Additionally, there are other scientific threats to moral responsibility posed by work in social psychology. For instance, it has been shown that the conscious reasons we tend to provide to explain our actions diverge from the actual causes of our actions and that, hence, our actions are often much less transparent to ourselves than we might assume (Bargh and Chartrand 1999; Wilson 2004; Uhlmann and Cohen 2005; Nosek et al. 2007). These findings reveal how the influence of external stimuli and events in our immediate environment can shape our behavior, even if we lack awareness of such influence. They also show how often our decisions and the resulting behaviors are driven by unconscious processes and cognitive biases (Kahneman 2011).

[5]The objection of several neuroscientists about the experimental setup notwithstanding, this has been the standard-interpretation for many years. For an alternative interpretation that conceives of Libet's readiness potential as resulting from spontaneous subthreshold fluctuations in neural activity that build up to form the readiness potential and that has no implications for moral responsibility see Schurger et al. 2012.

Even the weaker thesis that only the brain is a determinate system is not supported by the current evidence. While it is certainly correct that many brain processes on the cellular and subcellular levels are predictable, this claim does not hold on the level of larger neural networks. So while there is no route from neuroscientific findings to the truth of determinism (and hence no implication of either hard determinism or compatibilism), even neuroscientific findings on different levels of investigations do not definitely point to the interpretation of the brain as a predictable system. Therefore, it is hard to see how neuroscientific findings could undermine the incompatibilist conception of free will. Moreover, even neuroscientists have come up with alternative interpretations of the Libet experiments.

However, even if the alleged challenge to incompatibilism and moral responsibility is philosophically unsuccessful,[6] some have argued on the basis of further empirical evidence that the folk-interpretation of moral responsibility presupposes dualism between the physical world and mental states and, hence, incompatibilism and a libertarian conception of free will (Murray and Nahmias 2014; Nadelhoffer 2014).[7] That supports the argument that neuroscientific findings demonstrating how we tend to misattribute our agency to irrelevant causes could nonetheless undermine trust in our social practice of holding each other morally responsible. Irrespective of a strong argument against the argument from scientific predictability, there is evidence that many people (in western countries) *think* that libertarian incompatibilism provides the correct framework for thinking about free agency, and that science in general shows that we live in a universe that is completely causally determined (Nahmias et al. 2005; Nichols and Knobe 2007). Therefore, when neuroscience progresses and the biological mechanisms underlying our mental life become more and more predictable, this might result in an enhanced skepticism about free will and moral responsibility, as people could conceive of predictability as evidence for the truth of determinism (Nadelhoffer 2011). However, it also has been argued that the picture looks different if one confronts people with more palpable scenarios of moral responsibility. In these cases, philosophical laypersons tend to ascribe moral responsibility for action (and hence free will) even if the action was clearly causally determined but is in line with the agent's conscious motives for action (Murray and Nahmias 2014). This result can be interpreted such that whether people believe in an incompatibilist or compatibilist theory of free will and ascribe moral responsibility in accordance with the one or the other theoretical account depends on biases and motivated reasoning that allow us to maintain belief in moral responsibility. People prefer to be inconsistent regarding the metaphysical implications of their responsibility ascriptions, rather than giving up belief in moral responsibility (Clark et al. 2019).

[6] Here I am only concerned with the possibility of making a case against indeterminism by reference to neuroscientific and psychological investigations. That is not to deny that there are other, more convincing arguments against this view (cf. Frankfurt 1969, 1971; Nagel 1988; Zimmerman 1997 for a compatibilist or Pereboom 1995, 2014 for a hard determinist response).

[7] Dualism is a position nowadays held by hardly any philosopher. However, experimental philosophers have evidence that it is held by many philosophical laypersons nonetheless (Nadelhoffer 2014).

Skeptical arguments about moral responsibility usually try in one way or another to undermine the idea that we are in control of our actions. However, while the threatening interpretation of neuroscientific experiments directly addresses this task by stipulating the truth of determinism and then showing that we are not aware of the real causes underlying our behavioral choices, more traditional skeptical arguments try to challenge the capacity of controlling the course of our action by reasoning from these challenges to the conclusion that neither compatibilism nor incompatibilism about free will are suitable positions (Levy 2011; Pereboom 2014; Strawson 1994). Therefore, the problem is not only that we are unaware of the motives determining our actions, but rather that we are not in control of the factors that determine our motives[8] and their execution.[9] That we sometimes adopt motives for actions because the will forming processes have been manipulated through external factors, or that whom we have become and what we want is rather the result of different social, psychological, and biological factors that were beyond our control, seems to count against our capacity for moral responsibility – at least from a libertarian perspective.

In sum, neuroscience and psychology will not and cannot establish the truth of determinism and thus challenge the incompatibilist version of free will and moral responsibility. One may, however, wonder whether they can contribute to making eliminativist or revisionary responses to the question of whether human beings can rationally hold each other responsible in the retrospective sense *seem* more plausible. The more we learn how our motives, desires, and personalities emerge from factors that are beyond our control, and the more we realize that we are not always aware of what causes our behavior, the more reason we have to be skeptical about being the author of our actions; or so one might think. What is more, the recent interpretation of the inconsistent results from experimental studies on folk intuitions about free will and moral responsibility support the thesis that ascriptions of free agency and moral responsibility are the result of a robust psychological bias (Clark et al. 2019). If this interpretation is correct, it would on the one hand reveal a belief in free will as illusory due to the conceptually inconsistent pattern of ascription. But, on the other hand, it could perhaps render the threat of these results comparatively harmless. At least one might question whether belief in the illusory character of free will and moral responsibility will be enough to overcome our robust psychological bias. It might turn out that practical implications of holding the belief that the concepts of free will and moral responsibility cannot be applied in a consistent way, will not change our moral practice of holding each other morally responsible in a reasonable manner. That is to say that the normative implications of the purported neuroscientific challenges to free will and moral responsibility do not depend

[8]Consider arguments about how one's personal history shapes one's current beliefs and desires (c.f. Strawson 1994).

[9]Consider arguments from moral luck, according to which lucky circumstances that are beyond our control could intervene in the execution of our intentions or lucky circumstances bring about our having or not having certain desires that lead to morally problematic actions (c.f. Levy 2011; Zimmerman 1997).

on the philosophically most plausible interpretation of the cited experiments, but on what is publicly accepted as their most plausible interpretation. Thus, neuroscience's normative implications in the context of free agency and moral responsibility are at best indirect: The neurosciences stimulate discussions about how to think about agency and moral responsibility, but they do not have anything decisive to say to solve the underlying metaphysical questions. And what is more: they do not have anything decisive to say about how to respond to the perceived inconsistency of our ascriptions of free will and retrospective moral responsibility.

1.3 Level II: Debunking Moral Truth

Action theory is but one area in which neuroscience could have implications for the way normative concepts are to be understood and applied. Another area is moral epistemology, the study of how individuals come to know what is morally right and wrong. As I have pointed out above, normative consequences might not only arise through neuroscientific explanations of how human beings fail to ascribe moral responsibility for action but also from explanations of how we fail to recognize what action is actually morally desirable. Hence, another move to defend the thesis that neuroscience has normative implications is the claim that neuroscience is relevant to understanding how human beings grasp what is right or wrong. In this paragraph, I will argue that on a second, epistemic level neuroscience can at best claim to have normative implications given the metaethical debates about cognitivism and naturalism would be settled.

The practice of judging actions as "morally" right or wrong has been intensely studied in moral psychology and social neuroscience (Greene et al. 2001; Greene et al. 2004; Moll et al. 2008; Prinz 2011; Kahane et al. 2012; Cameron et al. 2018; Decety and Cowell 2018). While most of the studies in these areas do not explicitly derive normative consequences from their results but rather aim at describing how people's moral judgments change depending on various biological and social factors, some of these researchers have tried to argue that their empirical findings indicate that certain kinds of moral judgments are unreliable and should, hence, be dismissed. Most notably, Joshua Greene has argued that this is the case for deontological moral judgments because unlike utilitarian moral judgments, they are processed in brain areas that are evolutionary old and functionally associated with fast and frugal cognitive heuristics (Greene 2003). Greene founds this claim on a dual-process theory of moral reasoning and on two famous empirical fMRI studies (Greene et al. 2001, 2004). In a groundbreaking paper analyzing and reconstructing Greene's argumentation, Selim Berker has argued that Greene's conclusion against deontological judgment must either rely on bad inferences (such as: characteristically deontological moral judgments are wrong, because they rely on emotional processing) or require him to subscribe to a substantive normative premise, which would render the neuroscientific evidence provided by his study irrelevant to his normative claim (Berker 2009).[10]

Another notable example of the role of empirical research in moral epistemology is the Knobe Effect. The Knobe Effect refers to the observation that human beings consider an action's consequence to be intentionally brought about if they judge the consequence to be morally bad, but consider an action's consequence to be an unintended side-effect, if they judge the consequence to be morally good (Knobe 2003). This is a stunning result because whether an action's consequence is morally good or bad is irrelevant to the question of whether the consequence has been intended by the agent or not. However, the Knobe Effect shows that many people seem to think the opposite. Many humans have a psychological bias to judge that an agent deserves praise (and should be held responsible) for the morally good but unintended side-effects of an action, but does not deserve blame (and to be held responsible) for the morally bad but equally unintended side-effects of an action. The Knobe Effect then seems to describe a systematic failure in the human capacity to correctly ascribe responsibility. Knobe has claimed that these empirical findings undermine the doctrine of double effect, a principle important in deontological ethics, which states that morally bad consequences of an action are excusable if they are unintended side-effects. As we have seen, however, the Knobe Effect shows that people are only able to judge that a consequence should count as an unintended side-effect, if they hold the effect to be morally good. That is to say: The doctrine of double effect is psychologically implausible because people never judge that morally bad consequences are just side-effects.

Discussion of claims such as Greene's and Knobe's have shown that appeal to normative implications from neuroscientific evidence does not aim at directly deriving a normative claim from an empirical description, but at discrediting certain modes of moral reasoning based on so-called epistemological debunking arguments (Berker 2009; Kahane 2011; Fraser 2014; Liao 2017). In using a debunking argument, one claims that a belief or judgment is unjustified because it was produced by an epistemically unreliable process. The idea behind this reasoning is, hence, that neuroscience and moral psychology can give an empirical explanation of how cognitive processes can bring about certain moral judgments and that this explanation somehow discredits the content of the moral judgment. For example, the judgment that it would be right to toss a man off a bridge in front of a fast-moving train in order to save five people working on the train-track (a case that is known as the Footbridge dilemma case in normative ethics and is used by Greene and colleagues

[10] Berker (2009) argues that the most charitable reading of Greene's argument would be that he left out a normative premise. His argument, than, could be reconstructed as such:

Descriptive premise: Deontological moral judgments are driven by emotions.
Normative Premise: Moral judgments driven by emotions are wrong.
Conclusion: Therefore, deontological moral judgments are wrong.

Then, however, there is no *normative* role to play for neuroscientific evidence because this kind of evidence is only relevant to the descriptive premise. But see (Kumar and Campbell 2012) for a critique of this reconstruction.

in their fMRI studies) would be discredited if it could be empirically shown that only persons affected by cognitive biases or make a mistake in logical reasoning would judge as such.

From an epistemic point of view, it seems questionable what exactly can be learned from debunking arguments (Hanson 2016; Königs 2018). Obviously debunking explanations do not provide positive evidence in favor of a certain hypothesis, but only negative evidence against an alternative hypothesis. Hence, they allow falsification but not confirmation of a hypothesis. Therefore, they do not allow Greene, for instance, to claim that utilitarian moral judgments are correct or generally reliable, but only that they are more justified than others because they are based on an epistemically reliable process. Of course, this still allows for utilitarian moral judgments to turn out unfounded compared to potential other more reliable forms of moral judgments. The more fundamental problem is then that claims about the truth or falsehood of a certain kind of moral judgments easily overstretch the epistemic force of a debunking argument.

A related and more general problem is that debunking arguments built upon causal explanations of moral judgments that are supposed not to track the moral truth. However, such explanations can easily be found for many different kinds of moral judgments and beliefs and could even be targeted at morality in general. For example, it could be argued that all moral judgements are caused by processes in the brain and that our repertoire of neuronal processes represents adaptations to situations in evolutionary history that are in many ways different to the situations we humans face when we make moral judgments today. On this premise, it seems that the processes that cause our moral judgments are inadequate to track the moral truth. The result is that it seems possible to find a discrediting causal explanation not only for almost any type of moral judgment but even for morality as a whole (Kahane 2011). This is a problem even regarding the debate between consequentialists and deontologists in normative ethics because there is empirical evidence that can be interpreted as discrediting either side of the dispute. While Greene has argued that the cognitive processes involved with deontological moral judgments in the Footbridge case would discredit deontological moral judgments (Greene et al. 2004), other studies have found that persons with injuries in the ventromedial prefrontal cortex are more likely to make a "characteristically utilitarian judgment" when confronted with the above case than a control group without such a lesion (Ciaramelli et al. 2007; Koenigs and Tranel 2007). Moreover, further studies found that persons with frontotemporal dementia also tend to make "characteristically utilitarian judgments" in the Footbridge dilemma case more often than a control group (Mendez et al. 2005). Also, the likelihood to make utilitarian judgments in the Footbridge case increases in healthy persons after triggering emotional responses by letting them watch 5 min of comedy (Valdesolo and DeSteno 2016). It has been argued that all these pathological or manipulative influences have a causal role in bringing about "characteristically utilitarian moral judgments" and that they undermine their reliability in the same way as the results of Greene's experiments undermined the reliability of "characteristically deontological moral judgments" (Liao 2017). So, even if one considers the debunking strategy valid in principle to distinguish reliable from unreliable moral judgments, there currently is evidence that discredits both kinds of moral judgment. Therefore, it seems currently impossible to declare one kind of moral judgment to be more reliable than another by reference to empirical evidence alone.[11]

Yet another, rather obvious problem of debunking explanations is that they are not logically compelling. A moral judgment can of course still accidentally be right, even though the process that led to its formation was epistemically unreliable. But even granted that epistemically unreliable processes lead to false moral judgments most of the time and that, hence, debunking arguments can be plausible in principle, this shows that those who apply debunking strategies have to make some controversial metaethical commitments. Consequently, the plausibility of any purported normative implication of neuroscience that is based on a debunking argument also hinges on the plausibility of these metaethical assumptions.

The most obvious metaethical commitment of those who seek to debunk moral judgments through neuroscientific evidence is the claim that some moral judgments are more apt than others, viz. that there is some criterion to decide between the aptness of different moral judgments. If neuroscientific evidence is supposed to reveal certain moral judgments to be unreliable and – as a consequence – unjustified,[12] it must generally be possible to name a criterion of aptness for moral judgments. This can either be achieved by assuming some form of cognitivism, viz. the idea that moral judgements can be true or false, or by referring to some form of non-cognitivist metaethics (such as quasi-realism or fictionalism) that maintains the possibility of referring to moral judgments as if they could be true or false.

Cognitivism is a controversial view in contemporary metaethics insofar as its advantages (viz. the accurate representation of the phenomenal objectivity of the moral ought) do not accommodate for its obvious disadvantages (viz. the difficulty to explain the motivational character of purely cognitive moral judgments). Hence, cognitivism is a challenge to those who conceive of moral judgments primarily as expressions of persons' attitudes towards non-moral facts but cannot easily make sense of the objectively binding character expressed in moral language. Those who wish to claim that moral judgments can be false, hence, commit to the thesis that moral judgments describe some kind of moral facts. If one opts for cognitivism, one therefore needs to have some account of the nature of such moral facts and a theory how such facts fit with the world that is described by the natural sciences.[13]

---

[11] Similar observations have led some authors to argue that there possibly are not any epistemically reliable causal processes underlying moral reasoning (cf. Joyce 2007). Whether this would discredit morality as a whole is a question beyond the scope of this article.

[12] I take the claim that certain moral judgments or kinds of moral reasoning are "unreliable" or "irrational" as implying that the actions labelled as such are at least extremely likely to be *false*. In labelling a judgment as unreliable, one claims a high probability that the judgment is wrong. Similarly, to say that a cognitive process is irrational is to claim that it is does not track the truth and should be considered as false.

[13] Note, that I take cognitivism as defined above as compatible with any theory of truth. This means that cognitivism implies some kind of moral realism only in the sense that it must demand the existence of some independent entity that functions as a truth-maker for moral judgments. This,

Quasi-Realism (Blackburn 2000) and fictionalism (Joyce 2007), on the other hand, are no less controversial. While they maintain that moral judgments are expressions of evaluative attitudes that do not contain propositional content, they also subscribe to the view that evaluative attitudes are projected onto the world such that it appears to us as if it contained moral properties. Accordingly, moral judgments also appear to refer to properties of the world and can be treated as if they were true or false. These theories promise to allow for the presentation of moral oughts as objective and to maintain at the same time the intuition that moral judgments are directly motivating by virtue of being expressions of evaluative attitudes. As other non-cognitivist approaches, however, quasi-realism and factionalism have difficulties responding to cases in which moral judgments do not express a speaker's evaluative attitude (e.g., "If X is bad, then Y is also bad"), and they need convincing argument to show how the purely evaluative character of our moral language can create intersubjective obligation.

In light of these complexities, the prospects of deriving insights in what kind of moral judgments are more apt than others or even of deciding long-standing normative disputes in normative ethics between utilitarians and deontologists seem low. First, on a theoretical level, the debunking strategy only allows us to discredit certain kinds of moral reasoning as "unreliable", but not to convincingly show how and why they are unjustified. Second, it seems doubtful whether neuroscientific evidence can be relevant for successful debunking, as it can be used to discredit opposing kinds of moral judgments and, hence, does not provide a plausible distinctive feature. Finally, debunkers of moral judgment make metaethical commitments of which it is unclear if and how they could keep them. In the next section, I will examine this last point a little closer, to see whether the possibility of deriving normative claims from debunking arguments inspired by evidence from neuroscience can gain a higher plausibility. As we saw, one metaethical commitment neuroscientific debunkers of morality can have is cognitivism. On a theoretical level, there are several possibilities to account for such an explanation.[14] As I will show in the subsequent section, however, many neuroscientists have more or less explicitly endorsed specific forms of moral realism that can be subsumed under the term metaethical naturalism.

---

however, does of course not imply that cognitivists must commit to a correspondence theory of truth or to forms of moral realism that conceive of truth-makers as natural facts (for further considerations on this point c.f. Fisher 2010). The restriction of the following discussion on a special form of moral realism (viz. metaethical naturalism) is therefore only due to my intention to make the case for the relevance of neuroscientific evidence (which I take to be evidence about how the world "actually" is) in the realm of moral judgments as strong as possible and to account for some claims about this relevance that have actually been raised in neuroethical and metaethical debates.

[14] I will restrict the following discussion to cognitivist theories that assume the objectivity of moral facts, even though the connection of cognitivism with fictionalism (c.f. Joyce 2007) and constructivism (c.f. Street 2006) about moral facts has recently been revived at least with regard to general evolutionary debunking arguments.

1.4 Level III: Naturalizing Moral Properties

Metaethical naturalism in the broad and very vague sense of an approach to metaethics, which intends to cohere with naturalism in metaphysics more generally, conceives of all facts as natural facts and understands natural facts as facts of the sort the natural sciences deal with. Similar views seem to more or less explicitly underlie ideas of various neuroscientists and psychologists who insist that science should be taken as the best guide to reveal what is true about the world and what is true not only about but also in morality (Casebeer 2003; Gazzaniga 2006; Harris 2014). It is, however, not an easy task to spell out what exactly the idea of metaethical naturalism amounts to beyond the claim that human beings – including their rational and moral capacities – should be understood as being entirely a part of the natural world. In this final section, I will, therefore, examine two forms of metaethical naturalism that would allow conceiving of normative properties as features of the natural world that could (at least in principle) be analyzed by neuroscientific investigations.

In contemporary metaethics, naturalism has primarily been understood against the backdrop of G.E. Moore's intuitionist challenge that it would constitute a "naturalistic fallacy"[15] to define the basic evaluative term "good" by any other predicate (Moore 1903). Moore famously argued that "good" is an undefinable term, because one can always challenge the identification of a predicate X with "good" with the question "but is X really good?" For him, it is an open question whether a property identified as "good" is truly good. Through this argument, Moore claimed to have shown that evaluative and hence also moral properties are independent of the properties of the natural world.[16] Accordingly, metaethical naturalism is conceived of as a view denying Moore's intuitionist agenda. The ethical naturalist must show that evaluative and moral terms such as "good" and "bad", "right" and "wrong" can be fully explained by reference to natural properties, viz. that they are nothing over and above the properties that make up the things in the natural world.

Contemporary forms of metaethical naturalism have used different theoretical strategies to respond to this task. All forms of metaethical naturalisms, however, are committed to moral realism, as they all seek in one way or another to logically deduce evaluative terms and normative consequences from natural facts about the world. Hence, all naturalist versions of moral realism take moral properties to be properties of the world that exist independently from any psychological state of the subject that makes a moral judgment.[17] In the following, I will outline two approaches

[15] It has been argued in many other places that the term "naturalistic fallacy" as used by Moore is misleading. For Moore does not only term the identification of "good" with natural properties like "pleasurable" or "desirable" a naturalistic fallacy but also the identification of "good" with metaphysical properties such as "willed by god".

[16] As Moore was an intuitionist and non-cognitivist, and also considered the identification of "good" with non-natural properties as wrong, this view does not imply that moral properties factually exist in a reality beyond natural facts but rather that they do not exist as facts at all.

[17] Besides these naturalisms, there are, of course, further non-metaphysical versions of moral realism that conceive of objective moral properties in a different way. For example, subjectivists con-

to metaethical naturalism in current moral theory and discuss how they relate to the assumptions of some (neuro)scientists that morality is completely grounded in the natural world and can be explored in the same way as the objects of the natural sciences.

The first version of metaethical naturalism that should be discussed with regard to the neuroscience of morality is a non-reductive approach. This is to say that it conceives of moral properties as supervening on lower-level natural properties by which they are constituted, but does not propose that moral properties are identical to the lower-level natural properties. The relation between moral and natural properties, therefore, is non-reductive. However, as constituted things are also nothing over and above their constituents, this kind of "naturalist" realism can still be said to ontologically reduce moral properties to their basal, i.e., natural properties (Sturgeon 2013). This view, which has also been termed Cornell Realism,[18] opens an interesting path for a scientific understanding of the nature of morality.

A central assumption of Cornell Realism is that there is no fundamental difference between learning from observations in ethics and learning from observation in the natural sciences. Just as one can test theories about the natural fact in the sciences, one can test theories about the moral facts that hold in ethics (Boyd 1988). The idea behind this apparently bold analogy is that the indirect methods used in the sciences for proving the existence of unobservable natural entities could be analogously used in ethics to show the existence of moral facts. In the sciences, it is common practice to infer from the presence of observable phenomena to the existence of unobservable phenomena. For example, given a theory of thermodynamics, one can infer the degree of molecular motion in a liquid from the dilatation of mercury in a thermometer. In indirect observation, this inference is justified because of the correct background theory, which is conversely confirmed by the directly observable phenomena. Roughly speaking Cornell Realists try to transfer this idea to the moral realm. We are justified in the belief that our moral judgments reliably represent the moral facts, because our moral judgments are justified by reasonable moral theory, and the moral theory is reasonable, because it is supported by our moral judgments.[19]

According to this view, moral judgments play an irreducible explanatory role in our theory-dependent descriptions of the natural world (Boyd 1988; Sturgeon 1988). This is also a familiar phenomenon in scientific explanation. Scientific fields such

---

ceptualize moral facts as psychological facts such that a moral judgment like "Breaking promises is morally wrong" is true, if the judging person is in a corresponding psychological state. Still, others conceive of moral properties as a special kind of facts, which can only be described by reference to socially acquired moral concepts (c.f. McDowell 1994).

[18] The term "Cornell Realism" is used to label a set of theories that was developed by different philosophers who taught or studied at Cornell University. The most notable theorists of this philosophical school are Richard Boyd, Nicholas Sturgeon, and David Brink.

[19] This seeming circularity shows that Cornell Realists also try to transfer their contextualist understanding of justification and their commitment to confirmation holism (cf. Quine 1951) from the sciences to the moral realm.

as biology and psychology make use of certain concepts to describe natural phenomena which resist being analyzed in physical terms. Examples include "organism" in biology or "mind" in psychology. Nonetheless, these concepts certainly describe natural entities or phenomena, and if they are irreplaceable in our best theories about the natural world, the use of these concepts is unproblematic. Analogously concepts such as "good" or "right" have an irreducible explanatory function in ethics. For instance, we could explain the natural phenomenon of a change of moral norms only by the non-natural phenomenon that some people have recognized that some of their moral beliefs were actually wrong. It is, hence, the wrongness of these moral beliefs that explains the natural phenomenon of a historical change in moral norms.

Some psychologists and philosophers have developed approaches to explaining the normativity of moral judgments in neuroscientific terms with reference to some version of Cornell Realism (Kumar 2015, 2016). On such accounts, moral judgments have an irreplaceable function in explanations of human reasoning and behavior and, as such, constitute a moral fact. As shown, for example, by the above mentioned Knobe Effect, moral judgments explain how human beings reason about intentional action (Kumar 2016). Moreover, according to Kumar (2015, 2016) there is a large body of empirical evidence in moral psychology that shows that a manifold of cognitive processes could not be explained without referring to the concept of "moral judgement". Moral judgements influence not only how humans ascribe intention and responsibility for action but also how they ascribe knowledge (Beebe and Buckwalter 2010) and mental states (Pettit and Knobe 2009) to other persons. They even influence what people identify as causal relations (Alicke 2000; Knobe 2010). In the framework of Cornell Realism, the impossibility to explain certain kinds of non-moral human behavior without applying the concept of moral judgment reveals that moral judgments are irreplaceable entities in the natural world. This does not mean that it would be possible to identify moral judgments with some neural correlate, nor that one could directly derive normative implications from neuroscientific investigations into moral behavior. Yet, it indicates that understanding morality as a part of the natural world and, hence, conceiving of moral properties as real and irreducible entities is possible. If moral judgments as Kumar claims (Kumar 2015) form a "natural psychological kind",[20] this should provide room for neuroscientist and philosophers of mind to investigate how these natural kinds are realized on a neuronal level. Interpreted in this way neuroscientists and psychologists in search of a "brain-based" approach to ethics possibly are not completely on the wrong track. However, it is a further question if conceiving of moral judgments as irreplaceable natural kinds does settle any interesting normative issues. If, for example, a reference to typical utilitarian moral judgments would have a higher

[20] A "natural kind" is a group of particulars that exists independently of human efforts to order objects in the world. They are classes of objects not made up by human convention. Science is usually assumed to be able to reveal these kinds. Chemical elements or elemental particles such as electrons or quarks are usually considered to be natural kinds. Some philosophers also assume that some mental states, such as beliefs or desires, can constitute natural kinds (c.f. Ellis 2001).

explanatory power for phenomena in the natural world than a reference to typical deontological judgments, this would only indicate that the former is more likely to be an instance of a natural kind. Nevertheless, that still would not tell much about whether it would be good to judge that way.

The second version of metaethical naturalism important to the current debates in metaethics and neuroethics is Neo-Aristotelianism (Anscombe 1958; Foot 2003; Hursthouse 1999; Nussbaum and Sen 2002). Neo-Aristotelian approaches refrain from the endeavor of analyzing moral properties as natural properties and instead start from the idea of species-typical life forms. The idea is to make sense of what is morally right and wrong by seeing moral evaluation as a continuation of descriptions of species-typical functioning. For example, to say that a specific individual is a good member of its species is to say that all its parts and operations contribute in ways characteristic of this species to the ends of survival and reproduction. For human beings, however, there are more dimensions to species-typical functioning than just survival and reproduction. Human beings are only thought to function well if they are free of pain and capable of enjoying pleasures of sorts typical to their species (Hursthouse 1999). Moreover, human beings are also thought to be social and rational beings by nature and can, therefore, only flourish, if they are endowed with characteristics that conduce also to these ends. There is, so to say, a specific human life form defined by a set of final ends of which rationality and the capacity for reason-based action plays a central role. Neo-Aristotelian philosophers conceive of a human life that is conducive to these ends as a good life (Foot 2003). The reverse conclusion is, of course, that human beings that do not live up to these ends behave in a defective way.

What is important about Neo-Aristotelian naturalism in comparison to Cornell Realism and other forms of metaethical naturalism is that it does not start from a scientific picture of the physical world in which peculiar entities such as mental states or moral facts need to be fitted, but that it sees humans as beings which are already part of the physical world and fulfill their natural capacities within this world.

Even though this could be seen as an advantage of Neo-Aristotelianism, this kind of metaethical naturalism has significant problems as a basis for an understanding of morality as a phenomenon that could be analyzed by the natural sciences. The most obvious problem of these approaches is that current natural science has no room for teleological explanations. There are no predefined ends of plants, animals or human beings in a world that is governed by the laws of physics and evolution. Moreover, from a metaethical point of view, it is questionable if a functional understanding of "goodness" within the limits of a certain life form can account for the objectivity and universality characteristic of moral norms. To understand what it is that makes a human being live according to its function does not indicate what it is that makes an action morally right or wrong.

This has not, however, deterred some neuroscientists to conceive of virtue ethics, which is Neo-Aristotelianism's corresponding normative ethical theory, as the theoretical approach to human morality that fits best with the evidence about the neurobiology of moral judgment (Casebeer 2003; Casebeer and Churchland 2003). The argument for this connection is simply that neurobiological networks involved in what has been labelled "moral judgment" in fMRI studies are widely distributed over different brain areas with very diverse functions. Because virtue ethics conceives of moral judgments as very complex and context sensitive decisions involving a broad range of cognitive and emotional capacities, it is thought to best fit with the current empirical data on what happens in the brain during moral judgment. This line of reasoning reveals, however, that the neuroscientists' theoretical fondness of virtue ethics has no real backing in Neo-Aristotelian metaethics. Virtue ethics is favored as a normative theory simply because its practical implications correlate best with empirical findings (Casebeer 2003). The idea is that humans should take virtue ethics as a normative guideline because it seems to be the most adequate normative theory with regard to the cognitive functions they seemingly apply in making "moral judgments". Hence, this idea fits primarily with the functionalism in Neo-Aristotelian naturalism, but not with its inherent teleology. It seeks to derive the adequate normative theory (and as a consequence the possibility to make normative judgments) not from the typical functioning of the human being, but from the typical functions of the human brain.

It should be obvious that there are several things wrong with this line of reasoning: First, one may question what exactly the typical function of the human brain is. While it might still be uncontroversial to say that a heart's proper function in an organism is to enable blood circulation (rather than making sounds, for example), it seems a much more difficult question to settle what the proper function of the human brain consists in. As the brain plays a functional role in almost any kind of human behavior, brain function, as such, is not a promising criterion to distinguish between adequate and inadequate moral reactions. To base a normative ethical theory on brain functioning, therefore, requires first to distinguish between typical and atypical brain functions in moral reasoning. I cannot see, however, how such a distinction could be drawn with regard to the brain – an organ whose structure and function is so severely shaped by the interaction with the environment – without referring to independent norms of health and disease or notions of normality present in a given society. One cannot define the typical function of anything without making reference to the aims these functions serve in a given environment. But referring to aims already presupposes that this aim is valuable, i.e., that it has normative properties. Therefore, the functionalistic approach to the nature of moral judgments seems to beg the question.[21]

Second, this view on human moral behavior does not provide very concrete moral orientation. It only singles out a certain type of normative ethical theory as being supported by a functional analysis of brain areas involved in moral reasoning. But that does not tell us much about how human beings should live their lives. In a very benevolent reading one might indeed take these findings as a description of how functional human beings reason about moral questions. This indication, however, might be at odds with the reasoning required to arrive at the morally right

[21] This said I'd like to emphasize that this is not a general critique of Neo-Aristotelianism in ethics. The only claim is that this theory does not fit well with a scientific world view, which claims that there are no final ends in nature.

conclusions. For functionality and morality can very easily come apart. This problem is familiar to all approaches in virtue ethics: A highly functional human being in an immoral society will be an immoral agent.

The problem with Neo-Aristotelianism's familiar strategy to define morality in terms of functionality is, however, that this approach does not get off the hook of Moore's open question argument (Moore 1903). One can easily imagine situations in which one might question whether that what was functionally good was also morally good. My being functional in "mafia-style moral reasoning" does not make it likely that my actions will be particularly good from an impartial moral point of view. Still, some neuroscientists might want to object that there is no such thing as an impartial moral point of view. For instance, Casebeer has claimed that moral theories evolve with brain function. He thinks that humans only consider virtue ethics a plausible moral theory because it fits well with the way their brains function. Were human brains largely different from the brains we have today, human beings would develop a different moral theory, which would seem plausible to them because it would fit better with the function of their brain. This argument, however, completely fails to capture the objective sense of moral judgments. If one considers an action morally wrong, it should count as wrong for all rational beings and not just as wrong for those whose brains are accidentally wired in certain ways.

Finally, adopting this crooked Neo-Aristotelian perspective would have further normative consequences. It conceives of failure to judge and act in a morally right way as an instance of neural dysfunction. This way it parallels moral misconduct with a kind of brain disease. This conditions a one-sided view on social deviation and criminal behavior as biological dysfunctions and tends to mask the social and other environmental factors that condition the development of the biological dysfunctions.

In sum, the discussion of these two forms of metaethical naturalism has shown two things. First, in principle, there are theoretical options to describe normative properties and moral facts as part of the natural world. The two approaches discussed do not, however, eliminate normative concepts from a scientific understanding of the world entirely. Instead, they integrate them either by showing their explanatory value as concepts in our best scientific theories about the natural world or by deriving them from a functionalist understanding of the human life form. Second, while both help to theorize the nature of social and moral norms, it is not clear how the validity of concrete moral norms could be derived from both these kinds of metaethical naturalism. It seems that on this level again, normative implications are at best restrictions. At best, metaethical naturalisms can show that certain concrete moral judgments are implausible or do not hold, because they do not have any explanatory value or because they thwart the functions of the human life form. So even though these theories may have an important and interesting role to play for explaining what morality is and why human beings possess it, they still lack the potential to provide positive normative orientation.

1.5 Conclusion

In this paper, I argued that neuroscience does indeed have some limited normative implications. On the first level – the action-theoretic level – neuroscientific evidence has the potential to shape the way we use and apply normative concepts such as free agency, moral responsibility, or blameworthiness by providing evidence that people might be in control of their actions much less often than they think. While revealing new mechanisms of lacking self-control is indeed informative, this phenomenon is neither new in general nor specific of neuroscientific evidence. Neuroscience can provide empirical insights to inform moral practice just like other established disciplines such as social psychology. In the end, it is important how the neuroscientific evidence is interpreted and assessed with regard to its relevance for moral practice. For, to have normative implications neuroscientific evidence always needs to be accepted as relevant to the question at hand. But whether it is regarded as relevant, depends on factors beyond the realm of neuroscience itself.

I further argued that on the second, epistemic level neuroscience could only claim to have normative implications given metaethical commitments such as cognitivism and naturalism would be made explicit. This concerns first of all the question, whether neuroscientific debunking arguments are able to discredit normative intuitions, and if so, whether neuroscientific debunking can leave at least some of our moral intuitions untouched. For to argue that some type of moral judgment is less reliable than another, because it falls prey to a debunking argument, presupposes that one can rule out the possibility of a similar debunking argument against the defended moral view. As we saw, the prospects for restricting the coverage of neuroscientific debunking arguments are low.

Therefore, I finally examined two metaphysical theories that would allow conceiving of normative properties as features of the natural world that could, in principle, be analyzed by neuroscientific investigations. This step to the third level was important because it allowed assessing the prospects of "brain-based ethics" (Gazzaniga 2006), an ethics that grounds all its claims in the empirical knowledge we have about the functions of the human brain. While different forms of metaethical naturalism might indeed allow integrating normativity into a natural world that is entirely open for scientific investigations, it remained, however, unclear what would follow from these approaches normatively. While they explain how morality could be understood as a natural phenomenon, they do not directly allow us to derive specific normative judgments from this view. While Cornell Realism and Neo-Aristotelianism know to explain why and how normativity exists, both are poorly equipped for giving an account to what this discovery practically amounts.

What I did not consider in this paper was the effects media reports about neuroscientific debunking or about neurologically caused impairments of self-control have on the public discourse (this has been done elsewhere: O'Connor et al. 2012; Zimmerman and Racine 2012; Racine et al. 2017). As noted in the section on the agent-theoretic perspective, a socially accepted narrative can be the strongest game-changer even when the arguments backing the story are not completely sound from a philosophical or scientific point of view. Depending how much trust a community places in neuroscientific evidence, the echo this evidence finds in the public discourse might be the most important influence for changing the use of our normative concepts.

References

Alicke, M. 2000. Culpable Control and the Psychology of Blame. *Psychological Bulletin* 126 (4): 556–574.

Anscombe, G.E.M. 1958. Modern Moral Philosophy. *Philosophy* 33 (124): 1–19.

Bargh, J.A., and T.L. Chartrand. 1999. The Unbearable Automaticity of Being. *American Psychologist* 54 (7): 462–479.

Beebe, J., and W. Buckwalter. 2010. The Epistemic Side-Effect Effect. *Mind and Language* 25 (4): 474–498.

Berker, S. 2009. The Normative Insignificance of Neuroscience. *Philosophy & Public Affairs* 37 (4): 293–329.

Blackburn, S. 2000. *Ruling Passions. A Theory of Practical Reasoning*. Reprinted. Oxford: Clarendon Press.

Boyd, R. 1988. How to Be a Moral Realist. In *Essays on Moral Realism*, ed. G. Sayre-MacCord, 307–356. Ithaca: Cornell University Press.

Cacioppo, J.T. 2016. *Social Neuroscience*. Cambridge, MA: The MIT Press.

Cameron, C.D., et al. 2018. Damage to the Ventromedial Prefrontal Cortex is Associated with Impairments in Both Spontaneous and Deliberative Moral Judgments. *Neuropsychologia* 111: 261–268.

Casebeer, W.D. 2003. Moral Cognition and Its Neural Constituents. *Nature Reviews Neuroscience* 4 (10): 840–847.

Casebeer, W.D., and P.S. Churchland. 2003. The Neural Mechanisms of Moral Cognition. A Multi-Aspect Approach to Moral Judgment and Decision-Making. *Biology and Philosophy* 18 (1): 169–194.

Ciaramelli, E., et al. 2007. Selective Deficit in Personal Moral Judgment Following Damage to Ventromedial Prefrontal Cortex. *Social Cognitive and Affective Neuroscience* 2 (2): 84–92.

Clark, C.J., B.M. Winegard, and R.F. Baumeister. 2019. Forget the Folk: Moral Responsibility Preservation Motives and Other Conditions for Compatibilism. *Frontiers in Psychology* 10: 215.

Decety, J., and J.M. Cowell. 2018. Interpersonal Harm Aversion as a Necessary Foundation for Morality. A Developmental Neuroscience Perspective. *Development and Psychopathology* 30 (01): 153–164.

Ellis, B. 2001. *Scientific Essentialism*. Cambridge: Cambridge University Press.

Fisher, A. 2010. Cognitivism Without Realism. In *The Routledge Companion to Ethics*, ed. J. Skorupski, 346–355. New York/Routledge: Abingdon.

Foot, P. 2003. *Natural Goodness*. 1st ed. Oxford: Clarendon Press.

Frankfurt, H.G. 1969. Alternate Possibilities and Moral Responsibility. *The Journal of Philosophy* 66 (23): 829–839.

———. 1971. Freedom of the Will and the Concept of a Person. *The Journal of Philosophy* 68 (1): 5–20.

Fraser, B.J. 2014. Evolutionary Debunking Arguments and the Reliability of Moral Cognition. *Philosophical Studies* 168 (2): 457–473.

Gazzaniga, M.S. 2006. *The Ethical Brain. The Science of Our Moral Dilemmas*. 1st ed. New York: Harper Perennial.

Greene, J.D. 2003. From Neural 'Is' to Moral 'Ought': What are the Moral Implications of Neuroscientific Moral Psychology? *Nature Reviews Neuroscience* 4 (10): 846–850.

Greene, J.D., et al. 2001. An fMRI Investigation of Emotional Engagement in Moral Judgment. *Science (New York, N.Y.)* 293 (5537): 2105–2108.

———. 2004. The Neural Bases of Cognitive Conflict and Control in Moral Judgment. *Neuron* 44 (2): 389–400.

Haggard, P., and M. Eimer. 1999. On the Relation Between Brain Potentials and the Awareness of Voluntary Movements. *Experimental Brain Research* 126 (1): 128–133.

Hanson, L. 2016. The Real Problem with Evolutionary Debunking Arguments. *The Philosophical Quarterly* 90: pqw075.

Harris, S. 2014. *The Moral Landscape. How Science can Determine Human Values*. New York: Free Press.

Hursthouse, R. 1999. *On Virtue Ethics*. Oxford: Oxford University Press.

Joyce, R. 2007. *The Evolution of Morality*. Cambridge, MA: MIT.

Kahane, G. 2011. Evolutionary Debunking Arguments. *Noûs* 45 (1): 103–125.

Kahane, G., et al. 2012. The Neural Basis of Intuitive and Counterintuitive Moral Judgment. *Social Cognitive and Affective Neuroscience* 7 (4): 393–402.

Kahneman, D. 2011. *Thinking, Fast and Slow*. London: Lane.

Kitcher, P. 2014. *The Ethical Project*. Cambridge, MA: Harvard University Press.

Knobe, J. 2003. Intentional Action and Side Effects in Ordinary Language. *Analysis* 63 (3): 190–194.

———. 2010. Action Trees and Moral Judgment. *Topics in Cognitive Science* 2 (3): 555–578.

Koenigs, M., and D. Tranel. 2007. Irrational Economic Decision-Making after Ventromedial Prefrontal Damage. Evidence from the Ultimatum Game. *Journal of Neuroscience* 27 (4): 951–956.

Königs, P. 2018. Two Types of Debunking Arguments. *Philosophical Psychology* 31 (3): 383–402.

Kumar, V. 2015. Moral Judgment as a Natural Kind. *Philosophical Studies* 172 (11): 2887–2910.

———. 2016. The Empirical Identity of Moral Judgement. *The Philosophical Quarterly* 66 (265): 783–804.

Kumar, V., and R. Campbell. 2012. On the Normative Significance of Experimental Moral Psychology. *Philosophical Psychology* 25 (3): 311–330.

———. 2010. *Evolutionary Ethics*. Farnham/Burlington: Ashgate.

———. 2011. *Hard Luck: How Luck Undermines Freedom and Responsibility*. Oxford: Oxford University Press.

Liao, S.M. 2017. Neuroscience and Ethics. *Experimental Psychology* 64 (2): 82–92.

Libet, B., et al. 1983. Time of Conscious Intention to Act in Relation to Onset of Cerebral Activity (Readiness-Potential). The Unconscious Initiation of a Freely Voluntary Act. *Brain* 106 (Pt 3): 623–642.

Luethi, M.S., et al. 2016. Motivational Incentives Lead to a Strong Increase in Lateral Prefrontal Activity After Self-Control Exertion. *Social Cognitive and Affective Neuroscience* 11 (10): 1618–1626.

Mackie, J.L. 1985. *Ethics. Inventing Right and Wrong*. Harmondsworth: Penguin Books.

Matusall, S., M. Christen, and I. Kaufmann. 2011. The Emergence of Social Neuroscience as an Academic Discipline. In *The Oxford Handbook of Social Neuroscience*, ed. J. Decety and J.T. Cacioppo, 9–27. New York: Oxford University Press.

McDowell, J.H. 1994. *Mind and World*. Cambridge, MA: Harvard University Press.

Mendez, M.F., E. Anderson, and J.S. Shapira. 2005. An Investigation of Moral Judgment in Frontotemporal Dementia. *Cognitive and Behavioral Neurology* 18 (4): 193–197.

Moll, J., R. de Oliveira-Souza, and R. Zahn. 2008. The Neural Basis of Moral Cognition: Sentiments, Concepts, and Values. *Annals of the New York Academy of Sciences* 1124 (1): 161–180.

Moore, G.E. 1903. *Principia Ethica*. Cambridge: Cambridge University Press.

Murray, D., and E. Nahmias. 2014. Explaining Away Incompatibilist Intuitions. *Philosophy and Phenomenological Research* 88 (2): 434–467.

Nadelhoffer, T. 2011. The Threat of Shrinking Agency and Free Will Disillusionism. In *Conscious Will and Responsibility. A Tribute to Benjamin Libet*, Series in Neuroscience, Law, and Philosophy, ed. L. Nadel and W. Sinnott-Armstrong, 173–188. Oxford: Oxford University Press.

———. 2014. Dualism, Libertarianism, and Scientific Skepticism About Free Will. In *Moral Psychology, Volume 4. Free Will and Moral Responsibility*, ed. W. Sinnott-Armstrong, 209–216. Cambridge, MA: MIT Press.
Nagel, T. 1988. *Mortal Questions*. Cambridge: Cambridge University Press.
Nahmias, E., S. Morris, T. Nadelhoffer, and J. Turner. 2005. Surveying Freedom: Folk Intuitions About Free Will and Moral Responsibility. *Philosophical Psychology* 18 (5): 561–584.
Nichols, S., and J. Knobe. 2007. Moral Responsibility and Determinism: The Cognitive Science of Folk Intuitions. *Nous* 41 (4): 663–685.
Nosek, B.A., et al. 2007. Pervasiveness and Correlates of Implicit Attitudes and Stereotypes. *European Review of Social Psychology* 18 (1): 36–88.
Nussbaum, M.C., and A.K. Sen, eds. 2002. *The Quality of Life. A Study Prepared for the World Institute for Development Economics Research (WIDER) of the United Nations University*. Oxford: Clarendon Press.
O'Connor, C., G. Rees, and H. Joffe. 2012. Neuroscience in the Public Sphere. *Neuron* 74 (2): 220–226.
Paschke, L.M., et al. 2015. Motivation by Potential Gains and Losses Affects Control Processes via Different Mechanisms in the Attentional Network. *NeuroImage* 111: 549–561.
Pereboom, D. 1995. Determinism al Dente. *Noûs* 29 (1): 21–45.
———. 2014. *Free Will, Agency, and Meaning in Life*. Oxford: Oxford University Press.
Pettit, D., and J. Knobe. 2009. The Pervasive Impact of Moral Judgment. *Mind and Language* 24 (5): 586–604.
Prinz, J.J. 2011. Against Empathy. *The Southern Journal of Philosophy* 49: 214–233.
Quine, W.V.O. 1951. Two Dogmas of Epiricism. *The Philosophical Review* 60 (1): 20–43.
Racine, E., V. Nguyen, V. Saigle, and V. Dubljević. 2017. Media Portrayal of a Landmark Neuroscience Experiment on Free Will. *Science and Engineering Ethics* 23: 989–1017.
Ruse, M., and R.J. Richards. 2010. *Biology and the Foundations of Ethics*. Cambridge: Cambridge University Press.
Schurger, A., J.D. Sitt, and S. Dehaene. 2012. An Accumulator Model for Spontaneous Neural Activity Prior to Self-initiated Movement. *Proceedings of the National Academy of Sciences of the United States of America* 109 (42): E2904–E2913.
Singer, P. 2005. Ethics and Intuitions. *The Journal of Ethics* 9 (3–4): 331–352.
Spencer, H. 1897. *The Principles of Ethics, Vol. 1*. New York: D. Appleton Co.
Strawson, G. 1994. The Impossibility of Moral Responsibility. *Philosophical Studies* 75 (1–2): 5–24.
Street, S. 2006. A Darwinian Dilemma for Realist Theories of Value. *Philosophical Studies* 127 (1): 109–166.
Sturgeon, N.L. 1988. Moral Explanation. In *Essays on Moral Realism*, ed. G. Sayre-MacCord, 229–255. Ithaca: Cornell University Press.
———. 2013. Naturalism in Ethics. In *Concise Routledge Encyclopaedia of Philosophy*, ed. E. Craig, 615–617. Hoboken: Taylor and Francis.
Uhlmann, E.L., and G.L. Cohen. 2005. Constructed Criteria: Redefining Merit to Justify Discrimination. *Psychological Science* 16 (6): 474–480.
Valdesolo, P., and D. DeSteno. 2016. Manipulations of Emotional Context Shape Moral Judgment. *Psychological Science* 17 (6): 476–477.
Wilson, E.O. 1975. *Sociobiology. The New Synthesis*. Cambridge, MA: Harvard University Press.
Wilson, T.D. 2004. *Strangers to Ourselves. Discovering the Adaptive Unconscious*. Cambridge, MA/London: Belknap Press.
Zimmerman, M.J. 1997. Moral Responsibility and Ignorance. *Ethics* 107 (3): 410–426.
Zimmerman, E., and E. Racine. 2012. Ethical Issues in the Translation of Social Neuroscience. A Policy Analysis of Current Guidelines for Public Dialogue in Human Research. *Accountability in Research* 19 (1): 27–46.

Chapter 2
Moral Responsibility and Perceived Threats from Neuroscience

Myrto Mylopoulos

Abstract Neuroscience is offering daily insights into the inner workings of the brain and the biological basis of our decisions and actions. Such advances are exciting for many, but for others they bring along a sense of unease. This is because, insofar as neuroscience reveals our behavior to be the result of neural mechanisms, it seems for some to rule out the possibility that we are genuinely in charge of what we do, and that we can legitimately be held responsible for our actions. Thus, some have speculated that continued progress in neuroscience will fundamentally alter our conception of ourselves as free, responsible agents (e.g., Greene and Cohen 2004, p. 1775; Illes and Racine 2005, p. 14). In this chapter, I identify and evaluate three perceived threats to moral responsibility from neuroscience: (i) the threat from determinism, (ii) the threat from mechanism, and (iii) the threat from epiphenomenalism. I argue that worries for moral responsibility based on these perceived threats are ultimately unfounded. Neuroscientific findings may invite refinements and even important modifications of how we understand our own agency, but these adjustments would fall far short of any drastic revision and need not be interpreted in a negative light. To close, I suggest ways in which neuroscience, far from serving as a threat, may actually help us to enrich our understanding of ourselves as moral agents.

Keywords Moral responsibility · Neuroscience · Free will · Decision-making · Consciousness

---

M. Mylopoulos (✉)
Department of Philosophy and Department of Cognitive Science, Carleton University, Ottawa, ON, Canada

G. S. Holtzman, E. Hildt (eds.), *Does Neuroscience Have Normative Implications?*, The International Library of Ethics, Law and Technology 22,
https://doi.org/10.1007/978-3-030-56134-5_2

2.1 Introduction

Neuroscience is offering daily insights into the inner workings of the brain and the biological basis of our decisions and actions. Such advances are exciting for many, but for others they bring along a sense of unease. This is because, insofar as neuroscience reveals our behavior to be the result of neural mechanisms, it seems for some to rule out the possibility that we are genuinely in charge of what we do, and that we can legitimately be held responsible for our actions. Thus, some have speculated that continued progress in neuroscience will fundamentally alter our conception of ourselves as free, responsible agents (e.g., Greene and Cohen 2004, p. 1775; Illes and Racine 2005, p. 14).

In this chapter, I offer an analysis and evaluation of the implications of neuroscience for morally responsible agency, with the overarching aim of showing that worries of the sort just mentioned are ultimately unfounded. To be sure, neuroscientific findings may invite refinements and even important modifications of how we understand our own moral agency and the kind of control that underpins it, but these adjustments would fall far short of any drastic revisions, and need not be interpreted in a negative light. Or so I shall argue. To close, I will offer some more constructive remarks, by suggesting at least two ways in which neuroscience may actually help us to illuminate the psychological basis of moral responsibility.

To be clear, there is an important normative issue at stake in this discussion. This is because many assume that if an agent is not morally responsible for some action, then it is *unjustifiable* to hold them accountable for it. So, if neuroscience reveals that the conditions for being a morally responsible human agent are simply not met, then our current social practices of blame and punishment, reward and praise, ought not to continue, or at least would need to be dramatically altered.

In what follows, I identify and evaluate what I take to be three distinct perceived threats to moral responsibility from neuroscience: (i) the threat from determinism, (ii) the threat from mechanism, and (iii) the threat from epiphenomenalism (for related discussions on similar issues, see also Nahmias 2014 and Shepherd 2015). These seeming threats are intertwined in various ways. But dealing with them separately has its benefits, insofar as it allows us to more clearly identify the contribution of each to the overall sense that findings in neuroscience serve to undermine our everyday notions of agency and responsibility.[1] I turn now to this task.

[1] Here one might wonder why I am framing things in terms of the threat from neuroscience to moral responsibility, rather than free will, especially since many characterize free will as the control condition on moral responsibility, so a threat to the former would be a threat to the latter. But even if this is the case, neuroscience might reveal other threats to moral responsibility that do not proceed by undermining free will, and this is an important possibility to consider. In addition, in agreement with Schlosser (2013), I note that we care at least as much about moral responsibility as we do about free will, and yet much more attention has been devoted to drawing out implications for free will from neuroscience rather than drawing out those for moral responsibility, with the latter often only being taken up in connection with the former. (Schlosser himself restricts his

2.2 The Threat from Determinism

Determinism is often characterized as the view that the laws of nature, combined with the current state of the universe at a given time, together entail the state of the universe at any other time. If determinism is true, then our decisions and actions are links in a long causal chain of events that began well before we were born. Some think that this rules out the possibility that our actions are "up to us" in the way required for moral responsibility. (Not all agree, of course. So-called compatibilists about moral responsibility mentioned earlier, e.g., Fischer 1994, deny that this is a consequence of determinism.)

A number of different reasons have been offered for thinking this, but one view is that determinism rules out the possibility of being the "ultimate source" of one's actions, and this is what is required for genuine moral responsibility. The notion of an "ultimate source" is somewhat mysterious, but it is sometimes unpacked in terms of being an uncaused cause of one's behavior. Indeed, this is central to *agent causal* accounts of moral responsibility. These accounts hold that, as Chisholm writes, "[i]f we are responsible, [...] then we have a prerogative which some would attribute only to God: each of us, when we act, is a prime mover unmoved. In doing what we do, we cause certain events to happen, and nothing — or no one — causes us to cause those events to happen." (reprinted in Watson 1982, p. 32). Someone might hold, then that if our actions are determined, then we are not their ultimate sources in this way, and we thereby cannot be morally responsible for them. Another standard reason for accepting the incompatibility of moral responsibility and determinism, which does not require that we are uncaused causes of our actions, relates to the ability to do otherwise. Some have thought that, if determinism is true, then this ability is undermined, since what we do at any given time is entailed by the state of the universe and the laws of nature—there are no forking branches at our disposal in our future actions.

If one takes either of these views on board, then neuroscience might be thought to pose a threat from determinism insofar as neuroscientific results suggest that our decisions and actions are causally determined by prior brain activity. Support for this conclusion may be thought to come from widely discussed results of so-called subjective timing studies, which seek to establish the temporal relationship between unconscious neural activity and one's reported awareness of a decision to act. Results from these studies seem to indicate that what one decides to do can be predicted in advance on the basis of such activity. In particular, Libet et al.'s (1983) work in this area, based off the earlier pioneering work of Kornhuber and Deecke (1965), has been famously interpreted as showing that decisions to act are reliably preceded by a neural event that is thought to be the signature of action initiation—the so-called *Readiness Potential* (RP)—by an average of 350 ms. (I will further discuss this result later on.)

---

discussion to moral responsibility, but he does not centrally focus on what *neuroscience* implies for it, as I do here.)

More recent results suggest that the temporal gap between a neural decision to act and awareness of that decision can be even longer. Soon et al. (2008) asked participants to "freely decide" to press one of two buttons while watching a stream of letters that was refreshed every 500 ms. They were then presented with a screen of letters and told to identify the letter that was on display when they made their decision to act. Using fMRI and pattern-based decoders, Soon et al. first determined which brain regions encoded predictive information about the outcome of participants' motor decisions. They discovered that both the Supplementary Motor Area (SMA) and the pre-Supplementary-Motor-Area (pre-SMA) contained local patterns of fMRI activity that "encoded with high accuracy" specific motor decisions. Subsequently, they found that these patterns of fMRI activity were present well before the relevant decisions were consciously made, such that, adjusting for the slow time course of BOLD responses, the *outcome* of a decision could be predicted up to 10 s in advance from signals in the frontopolar cortex with about 60% accuracy (see also Bode et al. 2011).

Though these results may say some interesting things about the temporal relationship between consciousness and brain activity, they do not go any way towards actually establishing the truth of determinism. For one, it is important to highlight that in the Soon et al. study, the relevant decisions were only predictable about 10% above chance level. Perfect prediction was not possible, even as the time of the decision drew nearer. It is perfect prediction, however, that would provide some evidence for deterministic neural activity, and not just *any* degree of predictability—60% predictability still leaves wide open the possibility that indeterministic processes are at play. These results pose no threat from determinism.

In addition, given the current limits of our neuroimaging techniques, as well as EEG/EMG techniques, neuroscience tends to deal with very short time windows and simple decisions and actions (e.g., button presses). So even if there *were* at present strong evidence that deterministic processes lead up to our decisions in these very local and limited contexts, which there is not, this would fall far short of establishing that *all* our more complex actions and decisions are the result of deterministic causes. But it is this latter conclusion that must be established in order for the threat from determinism to be realized, since these are the decisions for which we tend to be held morally responsible.

Finally, as a last resort, even supposing that neuroscience were to give us reason to think that *at the level of the brain*, deterministic processes are pervasive, which again it does not, a further inference to the conclusion that the universe is deterministic is still not warranted. This is because neuroscience does not have the last word on this matter—that belongs to physics. Roskies (2006) makes an analogous point regarding the inference from indeterministic neural systems to a deterministic universe: "Because a deterministic system can radically diverge in its behavior depending on infinitesimal changes in initial conditions, no evidence for indeterminism at the level of neurons or regions of activation will have any bearing on the fundamental question of whether or not the universe is deterministic" (p. 421). Likewise, no evidence for determinism at this level can ultimately help us decide whether it is true of the fundamental level of physics.

So it seems that we do not have any reason to take seriously the threat from determinism. But this might not offer the reassurance that many are after, for they may take another more troubling threat to loom. I turn to this next.

2.3 The Threat from Mechanism

Perhaps what should be disconcerting about neuroscientific accounts of human action is not that they reveal deterministic processes at work, but that they reveal that our decisions and actions are the result of *mechanistic* processes. Mechanistic explanations are sometimes characterized as those that "propose to account for the behavior of a system in terms of the functions performed by its parts and the interactions between these parts" (Bechtel and Richardson 1993, p. 17). In the case of human agency, the "system" is the agent as a whole, and the "parts" can be construed as the various subsystems in the brain—e.g., the visual and motor system. So a main commitment of mechanistic explanation is that what an agent as a whole does can be exhaustively accounted for by a team of integrated subsystems working in concert towards some common set of goals or sub-goals.

Plenty of research in neuroscience is moving in the direction of mechanistic accounts of decision-making and action. The neuroscience of decision-making has come a long way in detailing the basic mechanisms underlying many of our simple decisions. A dominant view is that decisions in the brain, which can be viewed as commitments to a certain option among alternatives, are made by way of the accumulation of evidence to a specific threshold, or 'bound'. Consider how this is thought to work in perception. In a motion decision task, a group of sensory neurons in the middle temporal visual area (MT), selectively respond to moment-by-moment changes in the stimulus, selectively coding for leftward or rightward movement. Neurons in other areas (e.g., the lateral intraparietal area or LIP) then represent the accumulation of evidence for each choice from MT neurons over time, gradually increasing or decreasing their firing rate as evidence for each option is collected. Once a critical threshold in firing rate is reached, the decision process is terminated and an appropriate action is triggered (e.g., an eye movement in the correct direction). The specific level at which the decision threshold is set can be adjusted to accommodate various factors and is sensitive to cues indicating the success of the decision process (Shadlen and Roskies 2012).

Some argue that in the context of action, and specifically that which is operative in subjective timing studies like that of Libet et al., the decision of when to move is handled by a bounded accumulation model, except that in this case it is sensitive to internal noise and activity rather than features of external stimuli. Schurger et al. (2012) argue that when spontaneous fluctuations in neural activity happen to cross a critical threshold, the neural decision to act takes place. The important thing to note here is that it is the threshold-crossing neural activity that is to be properly identified with what Schurger et al. (2016, p. 77) refer to as the *neural* decision to act, since this is when the probability of the action occurring nears 1, and we can thereby construe the activity as reflecting a "commitment" to the action.

To support their interpretation of how decisions to move are arrived at in the Libet task, Schurger et al. conducted what they called the "Libetus Interruptus" task. They asked participants to perform Libet et al.'s (1983) original task, on each trial pressing a button spontaneously without any pre-planning. Participants were told that their waiting time might sometimes be interrupted by an audible "click", in which case they should press the button as quickly as possible. The prediction was that shorter response times should occur on trials in which spontaneous neuronal activity randomly happened to be closer to the decision threshold at the time the participants were interrupted. The resulting EEG data confirmed this prediction.

So far so good. But while bounded accumulation models are illuminating the underlying mechanisms of decision-making, Roskies (2018) observes the threat that we might confront if they end up being the basis for *all* of our decision-making:

> … the identification of neural activity with elements of this model can appear to pose a challenge to our intuitive understanding of decision-making as a process that is up to us, that is, an exercise of our high-level deliberative capacities, and an important nexus for the expression of free will. The realization of this abstract mathematical model in our brains may thus threaten to paint decision-making as a mechanistic, bottom-up process, unsuited to accommodating a notion of self-governance or top-down control that seems to describe our experiences of choosing, and that seems essential to the characterization of a responsible agent (p. 4).

The conmcern here can be put in terms of neuroscience entailing a reductive account of choosing, deciding, and initiation action. On such a reductive account, once we have given an appropriate account of the lower-level neural mechanisms involved in some decision, we have thereby given an account of these higher-level phenomena. But one might worry here that an agent is greater than the sum of their parts, and cannot simply be reduced to interactions among them; an explanation of human decision-making and action in terms of the neural mechanisms of the brain may necessarily leave out an active role for the agent, transforming them into a merely "passive" arena of events that occur "within" her.

To deepen the worry, as Roskies suggests towards the end of the quoted passage above, the purported problem can be further motivated by appeal to our subjective experiences of choosing, deciding, and acting. These experiences are rich and possess several interesting dimensions that will not concern us here (for a detailed discussion, see Mylopoulos and Shepherd 2020). But according to some, among the facts that they reveal is that your actions flow from *you*, the *agent*. This (admittedly somewhat elusive) aspect of the experience of acting has been referred to as the *sense of self as source*. Horgan (2012) describes it as follows:

> Suppose that you deliberately do something—say, holding up your right arm with palm forward and fingers together and extended vertically. What is your experience like? To begin with, there is of course the purely bodily-motion aspect of the phenomenology—the what-it's-like of being visually and kinesthetically presented with one's own right hand rising with palm forward and fingers together and pointing upward. But there is more to it than that, because you are experiencing this bodily motion not as something that is 'just happen-

> ing,' so to speak, but rather *as your own action*. You experience your arm, hand, and fingers as being moved *by you yourself*; this is the what-it's-like of *self as source*. (p. 64)

In other words, any time you act, you experience a robust sense of "top-down control", which involves you, yourself, initiating and guiding your action. If we take this phenomenological description on board, one can see why the threat from mechanism might be thought to be present, for a robust commitment to mechanism seems to suggest that this experience is non-veridical—there is no 'self' over and above the interacting systems in the brain revealed by neuroscience.

This way of motivating the threat can be further fueled by appeal to experiments that seem to reveal that the experience of self as source is *itself* the output of neural mechanisms and can be artificially simulated. For example, Fried et al. (1991) stimulated various cortical sites within the Supplementary Motor Area (SMA) of epileptic patients undergoing surgery and found that in some cases, even in the absence of any overt response, he induced in participants subjective experiences of having moved ("I feel my arm is moving") and "urges" to perform certain movements. More recently, Desmurget et al. (2009) stimulated parts of the parietal and premotor cortex in seven individuals undergoing open brain surgery, and once again found that in some cases the stimulation triggered conscious desires to move ("I felt a desire to lick my lips") and subjective experiences of having moved ("I moved my mouth, I talked, what did I say?") in the absence of any actual movement. So, the worry might go, if even our experiences of being the sources of our actions are the outputs of neural mechanisms, then how can we truly be the locus of control as they seem to suggest?

In response, we can first note that neuroscience alone does not have the resources to rule out a non-reductive account of choosing, deciding, and action initiation. Even if we were to discover all the neural correlates of our psychological states and processes, including our subjective experiences of agency, this would still leave room for them to be something over and above these neural correlates. There is no neuroscientific discovery that could determine that the "macro" phenomena (e.g., choices, decisions, and experiences thereof) described at the level of psychology are wholly reducible to the "micro" phenomena (e.g., neural patterns of activity) described at the level of neuroscience. This is a matter for theoretical inquiry and debate.

Moreover, it is unclear *why* we should deny that there is room for "top-down" agentive control within bounded accumulation models of decision-making. I submit that the main reason that it seems that there is not is that the kinds of decisions that neuroscientists are currently studying are low-level decisions of a sensory or motor variety that do not involve high degrees of complexity or flexibility. The decisions being modeled are often sub-personal in nature, and often within what would traditionally be characterized as encapsulated modules. It is not surprising, then, that they do not seem to leave room for agentive control, since they in fact are the types of decisions that tend to be automatic and below the threshold of conscious awareness. But these are patently not the types of decisions with which we are concerned in the moral domain. Here we care about such decisions as that of breaking a promise, giving to charity, or refraining from eating meat. These decisions are typically deliberate and cognitively global, involving as they often do mental processes that take into account a wide range of stored domain-general information. How this type of decision-making is implemented in the brain is simply not yet well understood by neuroscientists.

Suppose it turns out that what neuroscience eventually discovers is that even high-level decisions are the result of hierarchically organized bounded accumulation models, such that simple decisions are fed into more complex ones, and a wider range of factors are taken into account as "evidence" for a particular choice over its alternatives the further up the hierarchy one goes. Insofar as such accumulation models would involve sensitivity to information stored within the global architecture of the brain, it is difficult to see why this picture should be thought to fall prey to the concern that the agent is left out of the picture. Certainly the interactions that would be posited to explain various moral behaviors—those implicated for example in deliberation and empathy—would be such that they involve interactions among parts or subsystems, but these interactions would not be local in the way that those pertaining to simple visual or motor decisions are local. What else might constitute *bona fide* agentive control other than the operation of such a global system? What exactly has "disappeared", been "reduced", or "left out" here?

Here one might dig in one's heels and insist yet again that insofar as a reductive, mechanistic explanation of your behavior can be given, then *you* are not in charge. But without further unpacking this 'you', and clarifying exactly what it entails, and why it is incompatible with neuroscientific explanation of agency, this is not an answer to the questions just posed.

Now insofar as one remains unsatisfied with this attempt to dislodge the threat from mechanism, I suspect that this may be traced back to the sense that what is really left out of the picture here is a role for *consciousness*. After all, we have no first-person access to the brain mechanisms that are appealed to in neuroscientific explanations of our behavior. And for many, it is mystifying how such mechanisms can give rise to subjective experience in the first place (this is the so-called "hard problem" of consciousness as dubbed by Chalmers 1996). Thus, there seems to be a disconnect between purely mechanistic explanations of behavior, and any explanation that leaves room for consciousness to play a role. Making matters worse, many fear that consistent with this picture, neuroscience is revealing that consciousness is epiphenomenal. Epiphenomenalism about consciousness refers to the view that consciousness does not play any causal role in driving behavior: though it may *seem* to us as though we engage in conscious decision-making, and that it is in virtue of being conscious that those decisions come to guide our behavior, consciousness is really just "coming along for the ride"—it has no causal efficacy with respect to what we do.[2] I turn now to evaluate this worry and some of its purported implications.

[2] Note that, though I have suggested a connection between the threat from mechanism and the threat from epiphenomenalism, as a reviewer points out, both would hold regardless of whether determinism is true, and so are unrelated to the threat from determinism.

2.4 The Threat from Epiphenomenalism

Suppose that it is revealed to you that one of your closest friends is a zombie of the philosophical variety, so that while your friend is physically and functionally equivalent to you with respect to your psychological makeup, it turns out that they are not, and have never been, in any phenomenally conscious mental states. Nonetheless, they engage in all sorts of behavior that can readily be understood within a moral framework. They regularly donate money to charities, they volunteer at the local food bank, they never litter, they always keep their promises, they help elderly people cross the street, and so on. In light of their lack of state consciousness, can your friend really be said to be a moral agent? In other words, would we be justified in praising your friend for all of their good deeds?

Some would say 'no'. This is because, for them, consciousness and moral responsibility are inextricably linked. For example, the psychologist William Banks (2006) writes that: "We are not interested in unconscious freedom of the will, if there is such a thing, or unconscious volition … *From a legal or a moral standpoint, it is the conscious intention that counts in assigning blame or praise, and it is the conscious intention that the court or the moralizer tries to infer*" (p. 236, emphasis mine). Similarly, philosopher Timothy O'Connor (2009) claims that "[c]onscious awareness of one's motivations… is vital to the sort of freedom that consists in enjoying a significant moral autonomy" (p. 121). The bottom line seems to be that if one is not aware of one's motivations or decisions to act in particular ways, then one cannot be a full moral agent that is responsible for what they do.

In the past few decades, neuroscientific results have seemed to some to indicate that this picture is untenable. By now, several studies, including the Soon et al. (2008) study discussed earlier, have been marshalled in defense of the view that our conscious decisions to act arrive too late on the scene to be the true causes of our actions, and that it is unconscious brain activity that determines our behavior. On that basis, some have drawn skeptical conclusions regarding the extent to which we can trust in the veridicality of our experiences of agency (e.g., Wegner 2002), and even the possibility of free will and moral responsibility.

Much ink has been spilled pushing back on these skeptical conclusions, sometimes by way of pointing out various potential inaccuracies in the relevant empirical results (e.g., Banks and Isham 2009; Lau et al. 2007), and sometimes by way of disputing the interpretations of those results. Most convincingly, to my mind, Pacherie (2014) has recently argued that even if the results of subjective timing studies were to non-controversially show that we decide to act before we are aware of it, it is doubtful that this threatens conscious agency or moral responsibility, because exercises of agency go well beyond mere action initiation, which is the stage targeted by these studies. Rather, they encompass processes that occur before an action starts (planning and deliberation), as well as processes that occur after it has begun and as it unfolds ("online" action guidance). Our proximal decisions to initiate a particular action are but one part of this richer and more faithful understanding of human agency, and not a particularly important part with respect to our assignments of moral responsibility. When there *is* an emphasis on the importance of our intentions being conscious in a moral context, it makes the most sense to think of distal intentions that are the output of deliberation about what to do as being the type of intention at issue, not proximal intentions to initiate some action in the present moment that has already been decided upon. Subjective timing studies are silent on the role of consciousness within these broader aspects of agency.

Moreover, there is a general challenge that arises in interpreting neuroscientific results within the framework of moral psychology, which is directly relevant here. This is the challenge of how to map neural states and processes onto the familiar categories of folk psychology. We may know that certain neural events occur in the SMA prior to a conscious decision to act, but in the absence of some first-person awareness of such events, how do we determine what kinds of psychological states they correspond to? This gives us another reason to be skeptical of the dominant interpretation of results like those of Libet et al. (1983). In this vein, Mele (2007) has argued that the Readiness Potential (RP) is better interpreted as reflecting an unconscious urge or desire to A rather than an unconscious decision to A. On Mele's view, a decision to A is a momentary act of forming an intention to A. What distinguishes intentions and states like desires, urges, or inclinations, is that an intention involves some *commitment* on the part of the agent to an action plan. While desires are inputs to practical reasoning about what to do, intentions are the outputs of such reasoning. On Mele's interpretation, the RP does not reflect such a psychological commitment, but rather a prior, non-committal state that merely reflects one's being *inclined* to act.

Schurger et al.'s (2012) model discussed in the last section also supports such a view. On their model, the *buildup* of RP activity does not reflect a preparation or decision to move. Rather, it reflects the random ongoing "ebb and flow" of neural activity in the brain, and it is only when this activity crosses the critical threshold or bound that it can properly be characterized as the neural correlate of a decision to move. (For further discussion of this point, see Schurger et al. 2016.) Schurger et al.'s data are silent on the precise temporal relationship between the crossing of the neural threshold and the exact time that one is *aware* of a decision to move, so we cannot use this as evidence that the conscious decision and the neural decision coincide. But at the very least, claims to the effect that an unconscious decision to act occurs 350 ms before one is aware of it do not seem warranted. The same considerations also cast doubt on an interpretation of Soon et al.'s results wherein the brain "decides" to act significantly before you are aware of it.

So once again, we do not seem to have on the basis of neuroscientific findings alone, reason to take seriously the threat from epiphenomenalism.

2.5 Using Neuroscience to Illuminate Human Agency and Responsibility

If the arguments of the preceding sections are sound, then none of the perceived threats I have identified have any real bite. One might then be left with the question of whether neuroscience has *any* implications for moral responsibility after all. I think that it does, but that we have been looking in the wrong direction—for implications that are negative in nature. In this section, I identify two ways in which neuroscience can, rather, have a *positive* impact on our understanding of human agency and moral responsibility.

To start, I would like to push back on the kind of dismissal of the relevance of neuroscience for moral responsibility expressed in the following quotation from Gazzaniga and Steven (2005), who write:

> Neuroscience will never find the brain correlate of responsibility, because that is something we ascribe to people, not to brains. It is a moral value we demand of our fellow rule-following human beings. Brain scientists might be able to tell us what someone's mental state or brain condition is but cannot tell us when someone has too little control to be held responsible. The issue of responsibility is a social choice. According to neuroscience, no one person is more or less responsible than any other person for actions carried out. Responsibility is a social construct and exists in the rules of the society. It does not exist in the neuronal structures of the brain (p. 49).

Now there is some truth to what Gazzaniga and Steven say in this passage. It is true, for example, that neuroscience will "never find the brain correlate of responsibility"—indeed it is hard to see what this might mean, since responsibility is not itself a type of psychological state or experience. It is also true, strictly speaking, that brain scientists may not be able tell us "when someone has too little control to be held responsible". Still, it does not follow that neuroscientific findings are not importantly relevant to (i) the decision of whether someone should be held morally responsible in a given case, or to (ii) the identification of the general conditions that must be met in order for someone to be held morally responsible for their actions. And thus, normative implications of neuroscience for the question of when we ought to assign blame or praise to an agent may be overlooked by this narrow view.

On the question of whether or not one ought to be held morally responsible for some action, suppose that on a particular account of moral responsibility, one is morally responsible for some action A, only if, at the time of A-ing, they possess some psychological property P. And suppose that we know that some psychological property P is realized by brain area B. Here there is a way in which neuroscientific findings are potentially relevant for normative assessments of responsibility, since it is through neuroscientific tests and tools, including brain scans, that we might determine whether the relevant brain areas are functioning properly, and thus, in part, whether the agent does in fact possess the relevant property.

Indeed, there is a growing trend in this direction. Consider, for example, the use of neuroscience in the U.S. courtroom to argue for the lightening of a criminal sentence or to help assess the competency of a defendant, both by way of determining that some part of the brain that supports some psychological property relevant for moral responsibility, e.g., impulse control, has been significantly compromised. Thus, a recent study found that between the years 2005 and 2012, the number of judicial opinions appealing to neuroscientific evidence (e.g., history of brain damage or trauma, brain imaging) increased from approximately 100 to approximately 250–300 (Farahany 2016).

The utilization of neuroscience for such purposes is not without practical hurdles to clear. For instance, it may not be possible to determine with any certainty whether, in a particular case, someone did or did not perform some action *because of* an existing neuropsychological abnormality rather than some other cause. And we are a long way from understanding the neural basis for a host of psychological properties that are presumably relevant for moral responsibility, especially those related to domain-general reasoning and problem-solving. Still, there is promise, especially as neuroscientific measuring tools and tests become more reliable and accurate, for neuroscientific data to at least *contribute* to overall assessments of moral responsibility.

Similarly, on the question of which general conditions must be satisfied in order for one to be considered a morally responsible agent, neuroscience might help reveal the functional role of certain psychological properties, which is in turn important for determining whether their possession ought to be included as necessary conditions for moral responsibility. A helpful illustration of neuroscience at work in this way, as it relates to the psychological property of being conscious and whether or not it enables certain capacities that are themselves thought to be required for moral responsibility, comes from recent work by Levy (2014a, b). In his book, *Consciousness and Moral Responsibility*, Levy (2014b) puts forward what he calls the Consciousness Thesis (CT), which is that "[c]onsciousness of some of the facts that give our actions their moral significance is a necessary condition for moral responsibility" (p. 1). Here, the role reserved for consciousness is not that of enabling the direct control of behavior, in the sense that we entertained earlier, but in allowing an agent to assess and evaluate some aspect of what they are doing.

In particular, Levy's (2014b) focus is on consciousness of informational content. For Levy (2014a) "[t]he kind of consciousness at issue—awareness—is a state with contents of which the agent is aware" (p. 29). This awareness, in turn, requires that the state be personally available, i.e., that, "the agent is able to effortlessly and easily retrieve it for use in reasoning *and* it is online" (p. 33). In turn, for a state to be "effortlessly and easily retrievable" is for it to be such that a range of general cues would occurrently token that state, e.g., upon seeing your friend, having a thought about their recent loss of a loved one. For a state to be "online" is for it to be active in guiding an agent's present behavior, e.g., if this thought were to cause you to ask your friend how they are doing.

Let's look at a case appealed to by Levy (2014a) in which the consciousness condition is not met, and the relevant agent is, therefore, on his view, not morally responsible for their behavior. The case appeals centrally to psychological states that have been termed *implicit biases*. Implicit biases are often characterized as nonconscious psychological states that embody negative associations pertaining to members of certain social groups. These states are sometimes thought to be responsible for driving discriminatory behavior, ranging from "microaggressions" such as not making eye contact or sitting further away from a member of the targeted group, to large-scale behaviors like biased voting decisions (e.g., Knowles et al. 2010), biased hiring decisions (e.g., Uhlmann and Cohen 2005), and perceptual biases in identifying whether an item is a weapon or a tool (e.g., Payne 2001). Now consider the case of someone being on a hiring committee for a leadership position and selecting a male candidate over a female candidate, telling themselves that they do so on the basis of their qualifications, when in reality their final decision is the result of a sexist implicit bias that associates females with lack of leadership skills. Since they are not aware of the fact that gives their behavior moral significance, namely that it is guided by a sexist bias, on Levy's view, they are not directly morally responsible for this decision, despite its being a biased one.[3]

While cases like this might serve to motivate the CT, what is needed is a theoretical reason for thinking that consciousness is required for moral responsibility in the way that Levy suggests. And this, in turn, requires that we are clear about the general conditions that an agent must meet in order to be morally responsible, as well as how consciousness might help or enable an agent to meet those conditions. Philosophers have been characteristically keen to tell us just what these conditions are. Levy discusses both deep self (e.g., Smith 2008; Arplay 2002; Wolf 1990) and reasons-responsive (e.g., Fischer and Ravizza 1998) views of moral responsibility. In the interest of space, and because the case is perhaps more compelling, I will focus on reasons-responsive views here.

Reasons-responsive views of moral responsibility require that an agent exercise "guidance control" over their actions, which in turn requires that their actions are caused by a mechanism that is appropriately sensitive to reasons, where this is a function of what Fischer and Ravizza (1998) call "receptivity" and "reactivity". Receptivity is the ability to regularly recognize certain facts that would sufficiently weigh in favor of doing otherwise than the agent actually does. So, for example, consider a kleptomaniac who has compulsive urges to steal. If the kleptomaniac cannot recognize sufficient reason to do otherwise than they do, e.g., the fact that if they get caught they will get arrested, or the fact that the theft will cause significant distress for others, they are not reasons receptive. They need not recognize all such available facts, of course, but the ability to recognize at least one, or some subset, would suffice to make them receptive to reasons in the appropriate way. Reactivity, on the other hand, is the ability to *act* differently than the agent does on the basis of at least one, or some subset, of these facts. In some cases, the kleptomaniac might be able to recognize some of the reasons for not stealing, and yet be unable to act on them due to their compulsive urge to steal. Either way, we have a failure of reasons-responsiveness.

---

[3] It's worth noting that Levy allows that you are "indirectly" morally responsible, insofar as you can take measures to both weaken or extinguish your bias, and prevent it from manifesting in behavior.

Why think that consciousness is required for reasons-responsiveness? Levy (2014a) argues that reasons-responsiveness requires that we are aware of the facts that lead us to perform some action, and some fact pertaining to the moral significance of the action. This awareness of facts, in turn, requires that the relevant mental states carrying information pertaining to the facts be conscious. Why is this? Here is where neuroscience comes in. Levy subscribes to a leading neuroscientific account of consciousness, the Global Neuronal Workspace Theory (GNWT) (Baars 1997; Dehaene and Naccache 2001; Dehaene and Changeux 2011) of consciousness. On this view, consciousness is what ensures the personal availability of mental contents, by playing an integrative role. What this means is that when some informational content becomes conscious, it is made available to multiple regions in the brain for the purposes of planning and flexible control of action, verbal report, evaluation of information, and reasoning. As Levy (2014b) puts it: "When subjects are aware of information in this kind of way, its contents are available to a variety of consuming systems. Paradigmatically, the information is available to report (to self and others), at least in normal, awake subjects, but also available directly (that is, not in a manner mediated by report) to other systems" (p. 27).

How exactly is neuroscience contributing to this picture? Much of the motivation for the GNWT comes from findings in neuroscience that serve as the main evidence for thinking that consciousness plays the functional role that such theories ascribe to it. The "global workspace" within which information becomes available to consuming systems is thought to be a distributed neural system with "long range" neurons primarily residing in prefrontal and parietal cortex that serve to interconnect and coordinate modular subsystems by way of top-down recurrent feedback loops that serve to stabilize neural activation patterns (Dehaene and Changeux 2011). The activation of such a broad network of neurons has been associated with a number of important cognitive functions, such as conscious error detection as well as basic action awareness (Desmurget et al. 2009).

So in this case, neuroscience is crucial to a research program that aims to establish the nature of consciousness, and for some has important implications for the functional role that consciousness plays in supporting the kinds of capacities that one might take to be necessary for moral responsibility. Importantly, I am not here arguing that the GNWT is the best theory of consciousness, nor that the Consciousness Thesis (CT) is not without its problems. Perhaps there are good theoretical reasons to doubt both in the end. My main point here is that neuroscientific advances can be seen as a way forward for better understanding our own agency, rather than as a threat to our current conceptions.

2.6 Conclusion

In this chapter, I have identified three ways in which neuroscience is commonly supposed to be a threat to moral responsibility. I have argued that each seeming threat can be adequately dispelled. Thus, I strongly disagree with the suggestion that neuroscience will fundamentally alter our conception of ourselves as responsible agents in a negative way. Instead, I have proposed that neuroscience can actually contribute to a better understanding of moral responsibility. It can do so by helping us, first, to assess in particular cases whether an agent possesses specific psychological properties that are generally thought to be required for being held morally responsible, and second, to identify what those required psychological properties are, by helping us to understand what cognitive capacities they enable. If I am right, what we are left with is not cause for concern, but rather one for optimism regarding what the future of neuroscience can tell us about ourselves as moral agents.

References

Arplay, N. 2002. *Unprincipled Virtues: An Inquiry into Moral Agency*. New York: Oxford University Press.

Baars, B.J. 1997. *In the Theater of Consciousness: The Workspace of the Mind*. Oxford: Oxford University Press.

Banks, W. P. 2006. eds. S. Pockett, W. P. Banks, S. Gallagher. MIT Press.

Banks, W.P., and E.A. Isham. 2009. We Infer Rather Than Perceive the Moment We Decided to Act. *Psychological Science* 20: 17–21.

Bechtel, W., and R.E. Richardson. 1993. *Discovering Complexity: Decomposition and Localization as Strategies in Scientific Research*. Princeton: Princeton University Press.

Bode, S., A.H. He, C.S. Soon, R. Trampel, R. Turner, and J.-D. Haynes. 2011. Tracking the Unconscious Generation of Free Decisions Using Ultra High-Field fMRI. *PLoS One* 6 (6): 1–13.

Chalmers, D. 1996. *The Conscious Mind: In Search of a Fundamental Theory*. Oxford: Oxford University Press.

Dehaene, S., and J.P. Changeux. 2011. Experimental and Theoretical Approaches to Conscious Processing. *Neuron* 70 (2): 200–227.

Dehaene, S., and L. Naccache. 2001. Towards a Cognitive Neuroscience of Consciousness: Basic Evidence and a Workspace Framework. *Cognition* 79: 1–37.

Desmurget, M., K.T. Reilly, N. Richard, A. Szathmari, C. Mottolese, and A. Sirigu. 2009. Movement Intention After Parietal Cortex Stimulation in Humans. *Science* 324: 811–813.

Farahany, N.A. 2016. Neuroscience and Behavioral Genetics in US Crimincal Law: An Empirical Analysis. *Journal of Law and the Biosciences* 2 (3): 485–509.

Fischer, J.M. 1994. *The Metaphysics of Free Will: An Essay on Control*. Oxford: Blackwell.

Fischer, J.M., and M. Ravizza. 1998. *Responsibility and Control*. Cambridge: Cambridge University Press.

Fried, I., A. Katz, G. McCarthy, K.J. Sass, P. Williamson, S.S. Spencer, and D.D. Spencer. 1991. Functional Organization of Human Supplementary Motor Cortex Studied by Electrical Stimulation. *Journal of Neuroscience* 11 (11): 3656–3666.

Gazzaniga, M.S., and M.S. Steven. 2005. Neuroscience and the Law. *Scientific American Mind* 16: 42–49.

Greene, J., and J. Cohen. 2004. For the law, neuroscience changes nothing and everything. *Philosophical Transactions of the Royal Society of London. Series B, Biological Sciences* 359 (1451): 1775–1785.

Horgan, T. 2012. From agentive phenomenology to cognitive phenomenology: A guide for the perplexed. In *Cognitive phenomenology*, ed. T. Bayne and M. Montague. Oxford: Oxford University Press.

Illes, J., and E. Racine. 2005. Imaging or imagining? A Neuroethics challenge informed by genetics. *The American Journal of Bioethics* 5 (2): 5–18.
Knowles, E.D., B.S. Lowery, and R.L. Schaumberg. 2010. Racial Prejudice Predicts Opposition to Obama and His Health Care Reform Plan. *Journal of Experimental Social Psychology* 46: 420–423.
Kornhuber, H.H., and L. Deecke. 1965. Hirnpotentialänderungen bei Willkürbewegungen und passiven Bewegungen des Menschen: Bereitschaftspotential und reafferente Potentiale. [Changes in brain potentials with willful and passive movements in humans: the readiness potential and reafferent potentials]. *Pflügers Archiv* 284: 1–17. (in German).
Lau, H.C., R.D. Rogers, and R.E. Passingham. 2007. Manipulating the Experienced Onset of Intention After Action Execution. *Journal of Cognitive Neuroscience* 19 (1): 81–90.
Levy, N. 2014a. Consciousness, Implicit Attitudes and Moral Responsibility. *Noûs* 48 (1): 21–40.
———. 2014b. *Consciousness and Moral Responsibility*. New York: Oxford University Press.
Libet, B., C.A. Gleason, E.W. Wright, and D.K. Pearl. 1983. Time of Conscious Intention to Act in Relation to Onset of Cerebral Activity (Readiness-Potential). The Unconscious Initiation of a Freely Voluntary Act. *Brain* 106 (Pt 3): 623–642.
Mele, A. 2007. Free Will: Action Theory Meets Neuroscience. In *Intentionality, Deliberation, and Autonomy: The Action-Theoretic Basis of Practical Philosophy*, ed. C. Lumer and S. Nannini. London: Routledge Press.
Nahmias, E. 2014. Is Free Will an Illusion? Confronting Challenges from the Modern Mind Sciences. In *Moral Psychology (Vol. 4): Free Will and Moral Responsibility*, ed. W. Sinnott-Armstrong. Cambridge, MA: MIT Press.
Mylopoulos, M., and J. Shepherd. 2020. The experience of agency. In *The Oxford Handbook of the Philosophy of Consciousness*, ed. U. Kriegel. Oxford: Oxford University Press.
O'Connor, T. 2009. Degrees of freedom. *Philosophical Explorations* 12 (2): 119–125.
Pacherie, E. 2014. Can Conscious Agency be Saved? *Topoi* 33: 33–45.
Payne, B.K. 2001. Prejudice and Perception: The Role of Automatic and Controlled Processes in Misperceiving a Weapon. *Journal of Personality and Social Psychology* 81 (2): 181–192.
Roskies, A.L. 2006. Neuroscientific Challenges to Free Will and Moral Responsibility. *Trends in Cognitive Science* 10 (9): 419–423.
Roskies, A.L. 2018. Decision-Making and Self-Governing Systems. *Neuroethics* 11: 245–257.
Schlosser, M.E. 2013. Conscious Will, Reason-Responsiveness, and Moral Responsibility. *The Journal of Ethics* 17: 205–232.
Schurger, A., J.D. Sitt, and S. Dehaene. 2012. An Accumulator Model for Spontaneous Neural Activity Prior to Self-initiated Movement. *Proceedings of the National Academy of Science U S A* 109 (42): E2904–E2913.
Schurger, A., M. Mylopoulos, and D. Rosenthal. 2016. Neural Antecedents of Spontaneous Voluntary Movement: A New Perspective. *Trends in Cognitive Sciences* 20: 77–79.
Shadlen, M.N., and A.L. Roskies. 2012. The Neurobiology of Decision-Making and Responsibility: Reconciling Mechanism and Mindedness. *Frontiers in Neuroscience* 6: 1–12.
Shepherd, J. 2015. Scientific Challenges to Free Will and Moral Responsibility. *Philosophy Compass* 10 (3): 197–207.
Smith, A. 2008. Control, Responsibility, and Moral Assessment. *Philosophical Studies* 138: 367–392.
Soon, C.S., M. Brass, H.J. Heinze, and J.D. Haynes. 2008. Unconscious Determinants of Free Decisions in the Human Brain. *Nature Neuroscience* 11 (5): 543–545.
Uhlmann, E.L., and G.L. Cohen. 2005. Constructed Criteria: Redefining Merit to Justify Discrimination. *Psychological Science* 16 (6): 474–480.
Watson, G. 1982. *Free Will*. Oxford: Oxford University Press.
Wegner, D. 2002. *The Illusion of Conscious Will*. Cambridge, MA: Bradford Books.
Wolf, S. 1990. *Freedom and Reason*. Oxford: Oxford University Press.

Chapter 3
Lessons for Ethics from the Science of Pain

Jennifer Corns and Robert Cowan

Abstract Pain is ubiquitous. It is also surprisingly complex. In this chapter, we first provide a truncated overview of the neuroscience of pain. This overview reveals four surprising empirical discoveries about the nature of pain with relevance for ethics. In particular, we discuss the ways in which these discoveries both inform putative normative ethical principles concerning pain and illuminate metaethical debates concerning a realist, naturalist moral metaphysics, moral epistemology, and moral motivation. Taken as a whole, the chapter supports the surprising conclusion that the sciences have revealed that pain is less significant than one might have thought, while other neurological kinds may be more significance than has hitherto been recognised.

Keywords Pain · Affect · Motivation · Evaluativism · Motivational internalism

3.1 Introduction

Pain is ubiquitous. A wide range of situations may induce it; we may feel it when we break a bone, when we stare too long at the computer, when we eat – or especially when we drink – to excess. The pains we feel vary in intensity and exhibit profound variation in their felt qualities. A pain may be burning, searing, dull, rolling, gnawing, flashing, tearing, and much else besides.[1]

While pains are thus something that we typically feel, they are also something that we at least typically report as being located in our bodies. When I break a bone,

[1] The McGill Pain Questionnaire (Melzack 1975) is the most widely used tool for pain reporting in diagnostic contexts and includes more than 75 descriptors.

J. Corns
University of Glasgow, Glasgow, UK
e-mail: Jennifer.Corns@glasgow.ac.uk

R. Cowan (✉)
School of Humanities, University of Glasgow, Glasgow, UK
e-mail: Robert.Cowan@glasgow.ac.uk

G. S. Holtzman, E. Hildt (eds.), *Does Neuroscience Have Normative Implications?*, The International Library of Ethics, Law and Technology 22,
https://doi.org/10.1007/978-3-030-56134-5_3

I feel the pain *there*, in my arm where the damage has been done. When I stare too long at the computer I feel the pain just *here*, behind my eyes. When I eat or drink too much, the pain is not felt to be just anywhere, but is instead felt to be *in* my stomach and head. But what are these things, these pains, that I report as being located in my body? And how could pain be something located in my body, if it is also a felt experience of mine? This duality of pain – as something seemingly both felt and located – has given rise to a variety of puzzles about the nature of pain that have been of interest in both the humanities and the sciences.

As our rich body of everyday knowledge about pain attests, the complexity of pain does not end with this core duality. Pains appear to play a complex role in our biology, behaviour, and overall mental economy. Pains motivate action: when in pain, I am typically motivated to do something to make the pain stop, protect myself from further or worse pain, and to respond to any bodily damage that may have been the cause of my pain. Pains consume attention: when in pain, I am typically distracted from anything else that may be happening in my body, mind or environment – and the greater the pain's intensity, the more of my attention it appears to draw and retain. Pains are unpleasant: when in pain, I am typically in a state that feels crummy to be in and that I do not want to be in.

Different theories of pain have focused on different of pain's many complex roles, leading to rampant historical and contemporary disagreement about the nature of pain.[2] Philosophers have debated whether pain is a sensation, perception, emotion, or some combination thereof. In recent decades, conflicting neural accounts include a somatic perception theory of pain (Price 1999), pain as a homeostatic emotion (Craig 2003), the neuromatrix theory of pain (Melzack 2001), and more. These must eventually be reconciled with empirically-based psychological approaches, including biopsychosocial models (Gatchel et al. 2007). Pain scientists from across the disciplines continue to debate which, if any, neural pathways, areas, or activation patterns are constitutive or essential to pain.

Despite these theoretical controversies, we continue to take ourselves to have a rich store of everyday knowledge not only about pain's nature, but about its normativity. Accordingly, we seem to correctly predicate normative, including ethical, features of pain, i.e. many normative ethical claims about pain seem to be true. Pain is, in some sense, *bad*. Whatever – exactly – its nature turns out to be. As a result, we have a *pro tanto* reason to refrain from deliberately causing it.[3] Whatever –

---

[2] For discussion of the history of pain science, especially as relevant for philosophers, see Dallenbach (1939). For introductory overviews of the competing, contemporary, dominant views about the nature of pain in neuroscience, psychology, and philosophy, see the chapters in part 1 of the *Routledge Handbook of the Philosophy of Pain* (Corns 2017).

[3] A *pro tanto reason* for (or against) x-ing is a reason for (or against) x-ing which is defeasible but ineradicable. That is, it is a reason which can be overridden by competing considerations, but is not thereby cancelled. For example, for there to be a pro tanto reason to keep one's promises is for it to be the case that the fact that one's action would be the keeping of a promise always counts in favour of the action, even when, all-things-considered, there are overriding reasons not to keep one's promise (because, for instance, it would result in a catastrophe). This contrasts with *prima facie reasons*. A prima facie reason for (or against) x-ing is a reason for (or against) x-ing which is

exactly – it is. In general, we might initially think that normative ethical claims concerning pain, even those that require quantifying over pain, are insulated from scientific revelations. In what ways, we might sceptically ask, is pain science relevant for normative ethics?

When we turn to higher-order metaethical questions, i.e. questions concerning the nature of ethical thought and discourse and about the status of normative ethical claims, findings from pain inquiries may seem more straightforwardly germane, at least in principle. For example, scientific discoveries about the nature of pain are at least *prima facie* significant for various topics in moral metaphysics, moral psychology and epistemology. Despite these connections, there has been little engagement by metaethicists with contemporary developments in pain science. Thus, it seems useful to ask about the ways in which pain science can inform metaethics.

In what follows, we address these questions by highlighting a mere subset of the ways in which pain science, particularly neuroscience, is relevant for both normative ethics and metaethics. Regarding normative ethics, we argue that while experiences with negative affect, i.e. *unpleasant* experiences have normative significance, our best empirical theories about the nature of pain give us reason to think that pain, *as such,* is not normatively significant. Accordingly, certain ethical principles, e.g. those predicating wrongness of the deliberate causation of pain, are false. We trace three implications of the result of the normative insignificance of pain, as against unpleasant experiences, for normative theorising. Regarding metaethics, we similarly argue that contemporary pain and affective science (i) complicates certain kinds of "naturalistic" views about the metaphysics of ethical features; (ii) problematizes claims that pain and/or unpleasant experiences can ground normative or ethical knowledge; (iii) undermines certain views about moral motivation which appeal to the motivational profile of unpleasant experiences.

The roadmap is as follows. In Sect. 3.2, we offer a truncated overview of the neuroscience of pain focused upon four particular, and particularly surprising, empirical discoveries about the nature of pain. In Sect. 3.3, we focus on the relevance of these discoveries for normative ethics. In particular, we consider some normative ethical principles concerning pain and discuss some of the ways in which the discoveries presented in Sect. 3.2 are relevant to interpreting and applying those principles. In Sect. 3.4, we turn to metaethics. In particular, we illuminate the ways in which the complex nature of pain – as uncovered by pain science and exemplified by the discoveries expounded in Sect. 3.2 – are relevant to a realist, naturalist moral metaphysics, moral epistemology, and the debate about moral motivation. In Sect. 3.5, we conclude.

---

apparent "at first sight" but can be defeated and cancelled by other considerations. For example, for there to be a prima facie reason to trust one's memory is for there to be an initial presumption in favour of trusting one's memory, which can, however, be eliminated if one, for instance, discovered that one's memory was irredeemably faulty due to brain damage. In this paper we will be discussing normative principles concerning *pro tanto* moral reasons.

3.2 Neuroscientific Overview: The Surprising Complexity of Pain

To appreciate the state of the art in the contemporary neuroscience of pain, it is useful to begin with a glimpse into the recent past. The history of pain science may be helpfully understood as taking place against the backdrop of two cross-cutting debates. The first debate concerns whether pain is (1) a distinct type of feeling, or emotion (*the affective theory*), (2) a modality-specific sensation (*the sensory theory*) or (3) a sensation in any modality that crosses an intensity threshold (*the intensity theory*). The second debate concerns whether pain is a specific or convergent phenomenon. In the sense at issue, a 'specific' phenomenon is one that is realised by activities in biological structures dedicated to that phenomenon, whereas a 'convergent' phenomenon is one that is realised by the convergence of activities in multiple structures, none of which are dedicated to its occurrence.[4] Audition is a paradigmatically specific phenomenon, whereas cognition is a paradigmatically convergent phenomenon. These cross-cutting debates about pain's nature appear to have roots going back to Aristotle and the very beginning of pain science. By the mid-twentieth century, however, consensus seemed to be emerging that pain was a modality-specific sensation with dedicated areas, pathways, and structures. Though no such pain-specific pathways or areas had been identified, it was believed to be a mere matter of time. As a result, the sensory and specificity theories about pain dominated.

Beginning with the introduction of Melzack and Wall's gate-control theory in the 1960s, however (1965 and 1983), pain science – particularly the neuroscience of pain – began a dramatic transformation. Melzack and Wall presented evidence in favour of the novel, eponymous gating mechanism in the spinal cord being implicated in many (but crucially not all) pain experiences. Perhaps even more importantly, however, they presented evidence and arguments against both the sensory and specificity theories. The second historical debate – concerning whether pain was as specific or convergent phenomenon – was directly addressed. Pain, they argued, is a convergent phenomenon involving activity in multiple, dissociating areas and pathways – none of which are either necessary or sufficient for pain as reported. The first historical debate – about whether pain is an emotion, a modality-specific sensation, or a sensation crossing a certain intensity threshold – was revealed by the gate-control theory to be overly simplistic and to present something of a false trichotomy. There was no evidence for pain as a distinct modality and some evidence offered against it (contra the sensory theory), but it was also inappropriate to assume that pain was exclusively either an emotion (a la the affective theory) or a sensation (even if an intense one, a la the intensity theory). Pain, rather, was not only a convergent phenomenon, but a complex one, paradigmatically constituted by three explicitly identified dimensions (the terms for which are rough and ready descriptions of them): sensory-discriminative, affective-motivational, and

---

[4] For a more technical discussion of neural specificity as debated in the relevant sense, see Easter et al. (1985).

cognitive-evaluative. A paradigmatic pain involves activity across multiple parallel processing streams and has all three components. None of the activity in any of the identified pathways and areas implicated by each of these components is either necessary (required) or sufficient (enough) for pain. Just as crucially, and as discussed further below, these components (and the associated activity) doubly dissociate in surprising ways.[5]

In the wake of gate-control theory, all leading neuroscientific theories thus recognize the following surprising empirical discovery which has revolutionized pain science:

Discovery 1: Pain is a convergent phenomenon with multiple components which doubly dissociate.

Some of the discovered dissociations between the paradigmatic features of pain are more surprising than others and, as seen below, some are more relevant than others for normative and metaethical theorising. Three further discoveries are particularly worth stressing here, and we present them in order of increasing lingering controversy.

First, and least controversial, is the surprising discovery that pain as reported is poorly correlated with any specified type of bodily damage or disturbance. These poor correlations were one of Melzack and Wall's explicit motivations to identify, and encourage others to identify, previously neglected mechanisms and components of pain: convergent, complex theories can encompass novel mechanisms that may, if in distinct cases, explain the many cases of pain without identifiable bodily damage or disturbance, and the many cases of perceived bodily damage or disturbance without any reported pain. The previous specific, sensory theories had no explanation for these cases and patients suffering inexplicable pains were considered to be hallucinating or malingering. Pain science and medicine now almost universally acknowledge our next surprising discovery:

Discovery 2: Pains as reported and the registration of bodily damage (of any further specified type yet tested) are poorly correlated.

Third and equally important, is the surprising discovery that the sensory-discriminatory component of pain doubly dissociates from the affective-motivational component.[6] While both components are paradigmatic constituents of pain, the mechanisms involved in their realisation can, and often do, operate independently. This independent operation at the sub-personal, neural level is apparent at the personal level in cases of pains that are reported and located, but nonetheless claimed not to be unpleasant and which are unaccompanied by any behaviour taken to evidence unpleasantness, e.g. avoidance of the stimuli. Note that we can distinguish between the *primary affect* of pain – that is, the unpleasantness of the pain itself – and the *secondary affect* of pain – that is, the unpleasantness of the subsequent

[5] In general, two processes doubly dissociate when each can occur independently of the other.

[6] The study most often cited as initially establishing this surprising claim is Rainville et al. (1999), but many further investigations have been taken to support it and it is now widely accepted across pain science and medicine.

states that are caused by pain on a given occasion, e.g., anxiety or fear. Not only are there pains lacking any negative secondary affect – which may have been surprising enough – but the complex, convergent nature of pain is such that some pains appear to be lacking any negative primary affect. Some pains, that is, are not unpleasant.[7] We may summarize our third discovery accordingly:

Discovery 3: Pain does not always have negative primary affect, i.e. it is not always unpleasant.

While pains do not always have negative primary affect, they paradigmatically do, and our fourth and final surprising discovery focuses on this paradigmatic affective component. Negative affect, as noted above, was taken by Melzack and Wall to be essentially unified with the motivational role of pain: the affective-motivational dimension. Though the sensory-discriminatory component of pain may doubly dissociate from the affective-motivational component, the affective-motivational component was taken to be just that: a single, unified dimension, or component, of pain (and indeed any other unpleasant) experiences. More generally, many researchers in both the sciences and humanities have taken it for granted that negative affect motivates avoidance and positive affect motivates approach.

In recent decades, however, affective science has exploded and scientific investigations of the affective of pain and other experiences have advanced in ways that challenge this orthodoxy. Though this evidence is thus newer and remains more controversial than those discoveries noted so far, there is nonetheless now good reason to think that the affective and motivational dimensions of pain are, themselves, distinct and dissociable.[8] At the level of personal reports, distinctness is perhaps most clearly evidenced in cases of addiction: the addict is increasingly motivated to pursue activities and experiences with depreciating hedonic value, i.e. that are less and less pleasant. That is, as the motivation to seek the object of addiction increases, the pleasure in attaining the object decreases. When thinking directly about the neural mechanisms themselves, there is strong evidence for the independence of the activation, components, and activities of affective and motivational mechanisms.

Accordingly, we present a fourth and final surprising empirical discovery:

Discovery 4: The affective and motivational components of pain are distinct.

In what follows, we consider the implications of Discoveries 1–4 for both normative ethics and metaethics. Note that as these are all clearly contingent, empirical claims, we readily admit that they are open to challenge and to being subsequently overturned. Nonetheless, we take them to be well-supported and to represent important insights from the cutting-edge contemporary neuroscience of pain. Moreover,

[7] Such pains include some of those experienced while on morphine or other opiates, some chronic pains following lobotomy and leuchotomy, and pains of pain asymbolics (as against pain insensitives). For descriptions and initial references of empirical work concerning these, along with a central discussion of their philosophical relevance, see Grahek (2001).

[8] The work most clearly supporting this claim has been carried out by Berridge (see, for instance, Berridge 2004). For discussion of this work and its philosophical relevance see Corns (2014).

we think that consideration of their implications usefully illuminates the sorts of lessons that the neuroscience of pain may offer ethical theorising. The following two sections thus exemplify the relevance of neuroscience for ethics in principle, even if the particulars require alteration in response to subsequent discoveries.

3.3 Normative Ethics: The Surprising Moral Insignificance of Pain

For thinking about the relevance of the neuroscience of pain for normative ethics, we begin by noting that a number of seemingly plausible normative ethical principles – that would command quite widespread assent[9] – quantify over pain in uncontroversial formulations. Consider two:

Principle 1: It is pro tanto wrong to deliberately cause another creature pain.
Principle 2: There is always a pro tanto moral reason to minimize pain.

Prior to consideration of the pain science discussed in the previous section, we expect that these principles would strike the reader as highly credible. As far as first-order normative claims go, they seem relatively straightforward. Indeed, some might want to afford them the status of self-evident truths: knowable on the basis of proper understanding of them.[10]

However, in the light of empirical Discovery 1, we can see that there are a number of difficult questions that may be appropriately raised for the interpretation and application of these seemingly straightforward principles. Now that neuroscience has revealed that pain is a complex, convergent phenomenon, we might intelligibly ask which component or combination of components renders each of these principles true – if they are true. Which component(s) of pain makes it pro tanto wrong to cause it and which gives us a pro tanto reason to minimize it?

One natural thought is that the plausibility of both principles is underwritten by pain's negative affective component. It is natural, that is, to think that it is the nasty, unpleasant way that pains paradigmatically feel that makes it wrong to cause them and that gives us reasons to minimize them.[11]

If this is the only wrong-making feature of pain, then we should reject Principles 1–2 in favor of the following two principles:

---

[9] We later discuss utilitarian principles. But note that Principles 1 and 2 (and the other non-utilitarian principles we discuss) are likely to command more support among philosophers and non-philosophers than utilitarian ones.

[10] See, e.g., Audi (2004) for an account of self-evidence.

[11] Moreover, insofar as negative affect can occur non-consciously, it is plausibly *experienced* negative affect which is morally significant. For purposes of space, we set aside the question of whether negative affect is necessarily conscious. One of us doubts that it is, but we agree that non-conscious negative affect would at any rate be *less* morally significant than consciously experienced negative affect.

Principle 3: It is pro tanto wrong to deliberately cause another creature to experience negative affect.
Principle 4: There is always a pro tanto moral reason to minimize experienced negative affect.

These principles seem credible to us and they are clearly different than Principles 1–2, i.e. they have different truth conditions. Crucially, Discovery 3 states that not all pains have negative, primary affect, and there will accordingly be some pain experiences for which Principles 1–2 apply, while Principles 3–4 do not. If Principles 1–2 only seem plausible because Principles 3–4 are true, then it should only be wrong to cause pains that have negative affect. But is this the only reason that Principles 1–2 seem plausible?

To consider whether Principles 1–2 are supported solely by experienced negative affect, we should further disambiguate between two types of pains which both lack it: those involving bodily damage and those not involving bodily damage. Recall, as per Discovery 2, that bodily damage (and even its registration), is only poorly correlated with pain: there are many cases of pain without bodily damage (or its registration) and many cases of bodily damage (and even its registration) without pain. Nonetheless, pains paradigmatically *do* involve (the registration of) bodily damage and causing bodily damage might itself be a distinct wrong-making feature that supports the plausibility of Principles 1–2.

Accordingly, consider the following two plausible principles:

Principle 5: It is pro tanto morally wrong to deliberately cause bodily damage.
Principle 6: There is always a pro tanto moral reason to minimize bodily damage.

It might now seem natural to think that Principles 1–2 seem plausible only because Principles 3–4 and 5–6 are true. The only wrong-making or features of pain, we might think, are its paradigmatic negative affect and associated bodily damage. Take these away, and there is nothing of normative relevance left. If so, then we should reject Principles 1–2 and retain Principles 3–6.

Is this too strong? Consider a pain lacking both negative affect and bodily damage. If such a pain seems hard to imagine, note that the sensory-discriminatory component of a pain, even when present, often merely involves the signalling of high-threshold stimulation that is nonetheless not damaging stimulation. This is sometimes called 'potential damage.' Consider, for an everyday example, that when you put your hand on the stove, you feel pain prior to any actual damage being done to your hand. For a case of pain that is lacking both negative affect *and* bodily damage, you might now consider the pains of pain asymbolics.[12] Pain aysmbolics systematically report pains that are not unpleasant. Let us imagine, as there is good reason to do, that we take these reports at false value.[13] Imagine deliberately causing a pain to the asymbolic by pinching their arm hard enough to get the potential damage to be signalled, but not so hard that you actually cause any damage. The

[12] For a central discussion, see again Grahek (2001).
[13] For further discussion of the good reasons to take these reports at face value, see Corns (2014).

asymbolic has a pain experience and reports it as such. Is it pro tanto wrong to cause such a pain? Do you have a pro tanto reason to minimize any such pain?

We do not think the answers to these questions are obvious – which is just to say that, in the light of Discoveries 1–3, we do not think that Principles 1–2 are obvious. These principles seem obviously true if we think about paradigmatic pains for which all of the dissociating components of pains are present. If we take away the negative affect and any bodily damage, however, it is hard to see what feature of *the pain* would ground the wrongness of causing it or a reason to minimize it. To be sure, in our specifically imagined case, it is plausible that it is pro tanto wrong to pinch the arm of the asymbolic: interference of the asymbolic's autonomy and bodily integrity, for a start, are wrong-making features of the action. But, independently of such considerations, is there anything morally objectionable about causing *the pain* in that case?

The moral of the above line of thought is that by stripping the affective component of pain away, we arrive at a hedonically neutral (i.e. not at all unpleasant) sensation. If that sensation is *truly* hedonically neutral and its causation *truly* does not involve any bodily damage, then it is difficult to see why it is even pro tanto wrong to cause it or what reasons you could have (even pro tanto reasons) not to do so. It is, in summary, not credible to think that it is morally impermissible to cause a sensation *as such*. Consider any corresponding purported principles for causing visual or auditory sensations – if these are ever pro tanto wrong to deliberately cause, or if there are ever pro tanto reasons not to cause them, it is incredible to us that it is because of the sensations *as such*.

There may, of course, be some non-obvious reason(s) why we should accept not only Principles 3–6, but Principles 1–2. Pains, as noted in section I, typically consume and retain attention. We do not think it plausible, however, that the consumption and retention of attention, by itself, can do the requisite work. As a counterexample, consider orgasm, which we trust it is not even pro tanto wrong to cause and one does not have a reason (even pro tanto) not to cause. We are similarly skeptical, on similar grounds, about pains' paradigmatic involvement of motivation as reason to accept Principles 1–2. In general, notice that when evaluating any candidate wrong-making features of pain offered in support of Principles 1–2, we need to be careful to determine whether the wrong-making feature is merely a feature of a component or combination of components which pains may lack and which may be a feature of non-pains. It is, for instance, likely that it is something about the combination of the affective, evaluative, and cognitive features which make pain an effective – and exploitable – tool for learning. But it is far from obvious that pain, as such, is relevant here. Discoveries 1–2, again, allow us to see the difference. In sum, while there may be some non-obvious reasons to think pain as such is morally significant, we cannot think of any and know of nowhere they have been offered.

At this point, one might object that an experience isn't a *pain* experience unless it's morally significant. Maybe, that is, which normative principles are applicable to an experience is (at least part of) what determines whether or not that experience is a pain. Our everyday notion of pain, that is to say, is perhaps a normative notion.[14] Perhaps. But the discussion above highlights that it is far from obvious why we *should* take pain, as such, to have moral significance. In particular, once we realize that there are pains that do not involve any negative affect and that do not involve any damage, we may begin to wonder whether we should *revise* our notion of pain if it is, in fact, a normative notion. While paradigmatic pains have many wrong-making features, it is not clear why we should maintain that pains as such have any.

In general, then, we conclude that pain as such is not morally significant and that appearances to the contrary are best explained by the fact that many of the paradigmatic components of pain – singly and in combination – are. This conclusion has implications far beyond the status of Principles 1–2. We mention three.

First, one clear implication of the moral insignificance of pain is that the capacity to experience pain is then irrelevant not only for adult humans, but to the moral status of non-human, non-biological, or only partially developed creatures. Much contemporary and current discussion has regrettably focused on whether or not these kinds of creatures can feel pain; it has simply been assumed that feeling pain was either necessary or sufficient – or even both – for moral patiency. The question of whether non-human animals, such as fish or chickens, have the capacity to experience pain has seemed to many to be of great importance to whether or not we have any moral obligations towards these creatures.[15] Whether a fetus can feel pain has been taken to be directly relevant to normative and legal questions concerning abortion.[16] Whether we could make machines or artificial intelligence with the capacity to feel pain has raised troubling ethical quandaries for whether we should and what our obligations to such machines would be.[17] If the foregoing conclusions are correct, then these discussions – and their relevant deployments of the neuroscience of pain – are misplaced. While Principles 3–4 are plausible, Principles 1–2 are unmotivated.

This leads us to a second implication worth briefly mentioning concerning the status of classic utilitarian principles. Consider the following:

Principle 7: Actions are right in proportion as they tend to promote pleasure and wrong as they tend to promote pain.[18]

---

[14] See Dennett (1978) for an argument along these lines. Dennett, it is worth noting, here also argues for an eliminativism for pain, i.e. the claim that 'pain' never successful refers and so should be eliminated from everyday discourse. It seems to us that the revisionary move recommended in this text is also appropriate in response to his eliminativism, i.e. that we revise, without eliminating, our ordinary notion of pain in response to empirical discoveries. Space precludes further discussion.

[15] See, for example, Allen (2004).

[16] See, for example, Derbyshire et al. (1996).

[17] See, for example, Bostrom and Yudkowsky (2014).

[18] See, e.g., J.S. Mill: (1861, II 2) "The creed which accepts as the foundations of morals "utility" or the "greatest happiness principle" holds that actions are right in proportion as they tend to promote happiness; wrong as they tend to produce the reverse of happiness. By happiness is intended pleasure and the absence of pain; by unhappiness, pain and the privation of pleasure."

Even as we think Principles 1–2 should be rejected in favor of others (such as Principles 3–6), so we think that a classic or hedonic utilitarian should reject Principle 7 in favor of something like the following:

Principle 8: Actions are right in proportion as they tend to promote pleasure,[19] wrong as they tend to promote experienced negative affect.

Plausible utilitarian principles aren't, we think, best characterized as quantifying over pain, but are instead best understood as requiring that one minimize negative affect. We think it is clear that the relevant component which gives prima facie plausibility to Principle 7, and the component which utilitarians in fact care about, is the affective component. As Discoveries 1 and 3 make clear, Principles 7 and 8 are different principles. But there seems no reason that the utilitarian (qua utilitarian) should care about pains that are not unpleasant, and good reason to think that they should be equally concerned about unpleasant non-pains as they are about unpleasant pains. Even for the classic utilitarian, pain is morally insignificant.

As a third and final implication related to the previous two, recognizing the moral insignificance of pain helps illuminate the underappreciated significance of pains' paradigmatic components as they occur in non-pain experiences. In particularly, we suggest that negative affect–whenever it occurs – is morally significant and worthy of more consideration than it typically receives. Consider a recent trend in neuroscience, according to which some emotional experiences – particularly those that signal the loss or devaluation of desired relationships – are pains: the so-called 'social pains.'[20] As one of us has argued against this trend elsewhere (Corns 2015), the evidence is better explained by the fact that paradigmatic pains and the target emotional experiences both have negative affect. Nonetheless, we think that advocates of the social pain posit were right to stress the neglected and important neural similarities between unpleasant pain and unpleasant non-pains. These neural similarities, we now suggest, underwrite important normative similarity. We think that more moral weight should be accorded to other unpleasant non-pains, not only the so-called 'social pains' but other unpleasant experiences including paradigmatic confusion, stress, boredom, fear, tiredness, hunger, and so on. As with pain, the suggestion here is not that any of these unpleasant experiences, as such, are morally significant; rather, the suggestion is that the negative affect of these states is, ceteris paribus, as morally significant as the negative affect of pains. We needn't call an unpleasant state a 'social pain' or a 'mental pain' to increase its moral standing; rather, we should recognize that negative affect deserves moral consideration, though pain does not.

In summary, thinking about Discoveries 1–3 illuminates that for normative theorizing, we should shift our focus from pain to its morally significant components.

---

[19] For purposes here, we set aside the question of whether pleasure, like pain, is a complex, convergent phenomenon and whether the utilitarian should focus not on pleasure, but on positive affect. It is a matter of some controversy whether and to what degree pain and pleasure, or positive and negative affect, are symmetrical.

[20] See, for instance, Eisenberger and Lieberman (2004).

Morally insignificant pains occur when those components are absent, and those components are as significant when they occur in non-pain experiences as when they occur in pain experiences.

3.4 Metaethics: The Surprising Irrelevance of Pain

In this section, we briefly discuss the relevance of the neuroscience of pain, and particularly Discoveries 1–4, for three central areas of metaethics: moral metaphysics, moral motivation, and moral epistemology.

3.4.1 Moral Metaphysics

Many philosophers are attracted to some kind of moral realism, that is, the view that there are moral facts and properties, and that they are constitutively independent of the attitudes of actual or hypothetical agents.[21] A proponent of this view who thinks (plausibly!) that it is a fact that torturing children is wrong, thinks that this would be a fact even if all existing human agents believed otherwise. It would still be a fact even if it turned out that idealized versions of human agents, e.g. those freed of biases and equipped with *full* information, etc. would believe that torturing children is permissible. Among moral realists, some are also attracted to some kind of metaphysical naturalism, the view that there are only natural facts and properties. Although there is no consensus on what the distinction between natural and non-natural properties amounts to, in what follows we will assume the following plausible characterisation: a property is natural if and only if reference to that property would be useful for explanation or prediction in the generalizations of a completed or ideal science.[22]

A metaethical naturalist realist claims that moral properties are natural properties in the above sense. One way to develop this view would be to reduce moral properties to non-moral properties found in neuroscience. Such a naturalist, for instance, may identify moral badness, for instance, with a neurological kind. That is, they might claim that moral badness *is identical to* some property identified in our best neuorscience. Given its *prima facie* moral significance, as discussed above, pain might have been thought to be a plausible candidate for at least one such reductive base: the neurological kind to which badness can be reduced.

However, even as we have objected to the idea that pain as such is morally significant, so we think that the complexities of pain recently uncovered by

[21] See Huemer (2005) for discussion of this kind of view.

[22] For discussion on how to formulate naturalism and ethical naturalism, see Lenman (2006).

neuroscience constitute troubles for any reductive naturalist who wants to reduce a moral property like badness to pain.

One problem concerns the nature of pain itself: it is arguable that pain is not itself a natural kind, i.e. that reference to pain is not useful for explanation or prediction in the generalizations of a completed or ideal science. One lesson from recent pain science may be that the surprisingly complex, convergent nature of pain – as revealed in Sect. 3.2 – is such that reference to it is not useful for scientific explanation or prediction. Pain may not show up in a completed science. This may be the case if the dissociating components of pain involve mechanistic activity whose convergence, on any token occasion, is too idiosyncratic to support useful scientific explanations and predictions on other occasions. While our concept of pain may do well enough for everyday purposes, the complex, idiosyncratically convergent phenomenon that it picks out may not support scientific generalizations.[23] Distinct pains on distinct occasions may be so distinct, that a complete science will cease to proffer any generalizations across them; offering, instead, generalizations only about the mechanisms which dissociate and converge across these distinct occasions. If that's right, then pain could not provide a suitable reduction base for the ethical naturalist. While neuroscientific kinds are natural kinds, and so at least in principle could provide such a base, if pains are non-natural kinds, then they can't serve as the reductive base in a naturalistic moral metaphysics.

A second problem, also implied by the discussion in Sect. 3.3, concerns the moral properties that we are seeking to reduce: it is not clear that these moral properties are even *true* of pain, much less constituted by them. If pain as such is morally insignificant, then pain as such is not a promising candidate for a base property in a reductive ethical naturalist theory.

There appear, in sum, to be two relevant possibilities. First, one may maintain that badness is reducible to pain, but not – at least not yet – have thereby achieved a reduction of the ethical property to a natural kind. Pain would need to be further reduced to achieve naturalisation. Even if this problem could be overcome, however, we think the moral insignificance of pain problematizes this strategy. Second, one may give up reduction to pain, and maintain instead that badness is reducible to negative affect. We think this the more promising route. Both because we think negative affect as such, unlike pain, is plausibly morally significant, but also because the threats to the naturalness of pain based from its complex, convergent nature do not appear to arise for affect. Affective neurological kinds, e.g. affective mechanisms, areas, pathways, or neurochemicals, may present more plausible reductive possibilities. Thus is pain surprisingly irrelevant to the development of a successful naturalistic metaphysics of morals.

[23] See Corns (2012) for a sustained argument that pain is not a natural kind. Relatedly, see Roy and Wager (2017) for discussion of the idea, based directly on neuroscientific findings, that pain is a family resemblance, cluster kind, with no shared neurological nature.

3.4.2 Moral Epistemology

Ordinary moral agents seem to possess moral knowledge. For instance, we take ourselves to know that the perpetrators of the Holocaust were evil, or that torturing children is wrong. Moral epistemology focuses on whether and how such knowledge claims are true.

Pain initially appears relevant to moral epistemology in at least two ways. First, some of the moral knowledge that we take ourselves to possess concerns paradigmatic pains and their features. The normative principles discussed in the previous section are candidates for this kind of knowledge. Second, and more relevant for present purposes, moral philosophers as diverse as Plato and Frances Hutcheson[24] have been attracted to the idea that something like pain (and pleasure) plays a direct – that is, unmediated – role in grounding moral knowledge.[25]

One proposal is that pain grounds direct knowledge of the *moral* badness of bodily damage. There are many ways this proposal might be developed. One might be to appeal to a particular view about pain: Evaluativism. According to this popular view, (at least) paradigmatic instances of pain involve a representation of a bodily location being in a damaged state that is bad-for-the-subject.[26] On this view the representation of one's body being in a damaged state that is bad-for-the-subject accounts for paradigmatic pain's experiential unpleasantness. Perhaps, then, pain can provide its subject with direct knowledge that an instance of bodily damage is bad-for-the-subject, and perhaps this plays an indirect role in grounding our moral knowledge that causing bodily damage is pro tanto wrong.

As stated, this epistemological view – that pain can directly ground knowledge of *moral* badness – appears to identify moral badness with badness-for-the-subject. But one might object to this identification. Instead, moral badness is a kind of badness which is unrelativised to particular agents, i.e., it is badness *simpliciter*. For instance, one might think that the Holocaust was not only bad *for* all those who suffered as a result of it, but that it was also *morally bad* over-and-above this. If moral badness is badness simpliciter, one might think that pain could directly ground knowledge of moral badness only if it represented badness simpliciter. But on standard versions of Evaluativism, unpleasant pains only represent badness-for-a-subject. In response, one may attempt to argue that pains do sometimes represent badness simpliciter.[27] Alternatively, perhaps pain can indirectly ground moral knowledge – for instance, of general principles concerning the badness of bodily damage – even if it doesn't represent badness simpliciter.

A refinement of any such view, however, is necessary in light of Discovery 3: that pains do not always possess negative primary affect. Given this, and what might seem a reasonable assumption that pain would need to represent badness in order to

[24] See, e.g., Hutcheson (1725) and (1728).

[25] See Cowan (2017).

[26] See Bain (2017).

[27] But see Cutter and Tye (2011) for an argument against this.

ground knowledge of badness, it follows that only unpleasant pains will be capable of grounding moral knowledge. Moreover, there seems no reason to think that only pains represent the bodily-badness taken by the Evaluativist to explain pains' unpleasantness – hunger and tiredness are two other at least prima facie candidates. Even if these particular examples fail, we stress that it is the representation of badness that is the plausible candidate for grounding moral knowledge. If pains need to represent badness in order to ground knowledge of badness, then recognising that not all pains represent badness, may again lead us to conclude that pain, as such, is actually irrelevant – this time, epistemologically.

There is a further challenge for this kind of view. If we think that knowledge requires something like reliability – as many do – then Discovery 2 seems to problematize the idea that unpleasant pain can ground moral knowledge. Recall that it is widely accepted that there is a poor correlation between pain (unpleasant or not) and actual bodily damage. These include paradigmatic cases of unpleasant pains like headaches, migraines, lower-back pain, chronic pain, referred pain, and pain after healing. Given this, unpleasant pain itself does not appear to be – by itself – a reliable source of beliefs about whether one's body is damaged, and thus whether the damage is bad. This unreliability puts significant pressure on the claim that (even unpleasant) pain can ground moral knowledge.

Consider, briefly, an alternative epistemological proposal: pain experience grounds direct knowledge of the badness of pain experience. In light of Discovery 3, and the reasonable assumption that only unpleasant pains are bad, this proposal should only apply to cases of unpleasant pain. That is, it should be claimed that only unpleasant pains could ground direct knowledge of their badness. Note, however, that none of Discoveries 1, 2, and 4 appear to have any straightforward implications for this view. If unpleasant pains really are bad, then perhaps unpleasant pain episodes are a reliable source of beliefs about such badness. That is, perhaps subjects can reliably form true beliefs about the badness of their unpleasant pains on the basis of their experiences. Fully assessing this view, however, will require philosophical work on the nature of pain, including, crucially, the question of what (if any) evaluative content it possesses.

3.4.3 Moral Motivation

As evidenced by our own discussion in Sect. 3.2, affect has become an object of increasing scientific interest. Indeed, inquiries into the nature of affect have burgeoned into what is now considered a distinct science, *affective* science. It is perhaps unsurprising, then, that affect has become of interest across a number of philosophical debates.

An important example of this is the debate about moral motivation. It is agreed on all sides that moral judgment – that is, the mental attitude expressed in moral utterances – appears to have an intimate relationship with motivation. For instance, if I were to say to you that eating meat is a serious moral wrong, but then order a veal burger without hesitation, your initial thought would likely be that I don't really hold the moral view that I had uttered.

There is, however, disagreement concerning the precise nature of the connection between judgment and motivation. Roughly, Moral Motivational Internalists think that there is some necessary conceptual or metaphysical connection between moral judgment and motivation (perhaps owing to the fact that moral judgments are intrinsically motivating), while Motivational Externalists deny this, claiming that there is a merely contingent connection. To illustrate, consider again the veal example. About this Internalists will claim that, unless the agent was *somewhat* motivated to refrain from eating meat (perhaps overridden by their liking its taste), then they haven't made a genuine moral judgment. Externalists deny such an implication. Perhaps you make a genuine moral judgment but simply don't *care* about doing the right thing. Both sides have plausible arguments at their disposal, but neither seems to have a decisive advantage.[28]

Recently, however, some ethicists have thought we can make progress in the moral motivation debate by paying attention to the motivational profile of affective episodes, i.e. mental episodes that are paradigmatically pleasant or unpleasant, which allegedly play either a causal or constitutive role in moral judgements. For example, Jesse Prinz[29] has recently defended an affective version of Motivational Internalism on the basis of what he calls *Sentimentalism* about moral judgments, i.e., the view that all genuine moral judgments are constituted by emotions. Elsewhere, theorists like Kauppinen (2013) and Zagzebski (2003) have defended the view that there is a plurality of kinds of moral thoughts, so-called *Moral Thought Pluralism*, and that among these are a class of moral judgments which are (at least) intimately connected to emotions. Kauppinen thinks that some moral judgements are moral intuitions, and that these are constituted by emotions. Zagzebski thinks that the most developmentally and explanatorily basic moral judgments deploy thick concepts that are emotional. Call these appeals to affective states to advance the moral motivation debate the *Affective Appeal*.

The Affective Appeal is thought to advance the moral motivation debate since, as noted in Sect. 3.2, affect has traditionally been assumed to be essentially, or constitutively, tied to motivation. Attributing an affective component to all or some moral thoughts has thus been assumed by the aforementioned theorists to thereby explain why and how it is the case that there is a necessary connection between some or all moral thoughts (the affective ones) and motivation. Further, the Affective Appeal allegedly explains the necessary connection between some or all moral thought and motivation in a way that is both more empirically respectable and less theoretically controversial than non-affective Internalist theories.

However, we think that a general problem for all such versions of the Affective Appeal is Discovery 4: that the affective and motivational components of pain are distinct. If affect and motivation really are distinct and independent, then simply

---

[28] See also the chapter by Zarpentine, this volume.

[29] See his (2015).

attributing an affective component to a kind of moral thought is insufficient to ground a necessary connection between moral thoughts and motivation. Crucially, some states may motivate without being affective and some states may be affective without motivating. The proponent of the Affective Appeal will thus need to amend or supplement their view in order to explain why, as they claim, there is some kind of necessary connection between moral thought and motivation.[30]

The general lesson from Discovery 4 for debates about moral motivation is that affect is, by itself, insufficient to explain a necessary connection between moral judgment and motivation. If moral judgements motivate in one of these Internalist ways, it is not merely because they are affective. To explain moral motivation, or to advance the debate between Motivational Internalists and Externalists, we need to look elsewhere. Despite its relevance vis-à-vis other metaethical and normative views, affect is here surprisingly irrelevant.

3.5 Conclusion

In everyday life, we take ourselves to have a rich body of knowledge about pain – both descriptive and normative. Pain science, however, has undergone dramatic advance, uncovering surprising discoveries about its complex nature, and we have here argued that these discoveries entail surprises for normative and metaethical theorizing. In particular, the complexity of pain as revealed by neuroscience shows us that pain, as such, is less significant and relevant than philosophical orthodoxy has thought. More positively, it suggests that other neurological kinds may be more worthy of moral consideration and more important for metaethics than we have collectively realized.

Even if the particular discoveries canvassed here are subsequently overturned, we trust that this chapter evidences the kinds of important lessons for ethics that are promised by the science of pain. As both the sciences and humanities advance, it is our hope that they continue to inform one another so that our theories, and our lives, improve.

References

Allen, C. 2004. Animal Pain. *Nous* 38: 617–643.
Audi, R. 2004. *The Good in the Right: A Theory of Intuition and Intrinsic Value*. Princeton: Princeton University Press.

[30]As we argue in Corns and Cowan (2018), it is far from obvious that the Affective Appeal can straightforwardly ground the truth even of attenuated versions of Internalism, e.g. ones which posit a necessary connection between moral judgment and motivation only for *well-functioning* or *practically rational* agents.

Bain, D. 2017. Evaluativist Accounts of Pain's Unpleasantness. In *The Routledge Handbook of the Philosophy of Pain*, ed. J. Corns. London: Routledge.
Berridge, K. 2004. Motivation Concepts in Behavioral Neuroscience. *Physiology and Behavior* 81: 179–209.
Bostrom, N., and E. Yudkowsky. 2014. The Ethics of Artificial Intelligence. In *The Cambridge Handbook of Artificial Intelligence*, 316–334. Cambridge: Cambridge University Press.
Corns, J. 2012. *Pain is Not a Natural Kind*. City University of New York.
———. 2014. Unpleasantness, Motivational *oomph*, and Painfulness. *Mind and Language* 29 (2): 238–254.
———. 2015. The Social Pain Posit. *Australasian Journal of Philosophy* 93 (3): 561–582.
———., ed. 2017. *The Routledge Handbook of the Philosophy of Pain*. New York: Routledge.
Corns, J., and R. Cowan. 2018. *Moral Motivational Internalism and the Affective Appeal*. Manuscript Submitted for Publication.
Cowan, R. 2017. Pain and Justified Evaluative Belief. In *The Routledge Handbook of the Philosophy of Pain*, ed. J. Corns, 354–364. New York: Routledge.
Craig, A.D. 2003. A New View of Pain as a Homeostatic Emotion. *Trends in Neurosciences* 26 (6): 303–307.
Cutter, B., and M. Tye. 2011. Tracking Representationalism and the Painfulness of Pain. *Philosophical Issues* 21 (1): 90–109.
Dallenbach, K.M. 1939. Pain: History and Present Status. *American Journal of Psychology* 3: 331–347.
Dennett, D.C. 1978. Why You Can't Make a Computer that Feels Pain. In *Brainstorms*. Cambridge, MA: MIT Press.
Derbyshire, S.G., A. Furedi, V. Glover, N. Fisk, Z. Szawarski, A.R. Lloyd Thomas, and Maria Fitzgerald. 1996. Do Fetuses Feel Pain? *BMJ: British Medical Journal* 313 (7060): 795–798.
Easter, S.S., D. Purves, P. Rakic, and N.C. Spitzer. 1985. The Changing View of Neural Specificity. *Science* 230 (4725): 507–511.
Eisenberger, N.I., and M.D. Lieberman. 2004. Why Rejection Hurts: A Common Neural Alarm System for Physical and Social Pain. *Trends in Cognitive Sciences* 8 (7): 294–300.
Gatchel, R.J., Y.B. Peng, M.L. Peters, P.N. Fuchs, and D.C. Turk. 2007. The Biopsychosocial Approach to Chronic Pain: Scientific Advances and Future Directions. *Psychological Bulletin* 133 (4): 581–624.
Grahek, N. 2001. *Feeling Pain and Being in Pain*. Cambridge, MA: MIT Press.
Huemer, M. 2005. *Ethical Intuitionism*. New York: Palgrave Macmillan.
Hutcheson, F. 1725. *An Inquiry Concerning Moral Good and Evil*, in Raphael, D.D., *British Moralists* 1650–1800, Hackett.
———. 1728. *An Essay on the Nature and Conduct of the Passions, with Illustrations Upon the Moral Sense*, in Raphael, D.D., *British Moralists* 1650–1800, Hackett.
Kauppinen, A. 2013. A Humean Theory of Moral Intuition. *Canadian Journal of Philosophy* 43 (3): 360–381.
———. 2015. Intuition and Belief in Moral Motivation. In *Motivational Internalism*, ed. G. Björnsson et al. Oxford: Oxford University Press.
Lenman, J. 2006. *"Ethical Naturalism" in the Stanford Encyclopedia of Philosophy*. https://plato.stanford.edu/entries/naturalism-moral/.
Melzack, R. 1975. The McGill Pain Questionnaire: Major Properties and Scoring Methods. *Pain* 1 (3): 277–299.
———. 2001. Pain and the Neuromatrix in the Brain. *Journal of Dental Education* 65: 1378–1382.
Melzack, R., and P. Wall. 1965. Pain Mechanisms: A New Theory. *Science* 150 (3699): 971–979.
———. 1983. *The Challenge of Pain*. New York: Basic Books.
Price, D.D. 1999. *Psychological Mechanisms of Pain and Analgesia*. Seattle: IASP Press.
Rainville, P., B. Carrier, R.K. Hofbauer, M.C. Bushnell, and G.H. Duncan. 1999. Dissociation of Sensory and Affective Dimensions of Pain Using Hypnotic Modulation. *Pain* 82 (2): 159–571.

Roy, M., and Tor Wager. 2017. Neuromatrix Theory of Pain. In *The Routledge Handbook of the Philosophy of Pain*, ed. J. Corns, 87–88. New York: Routledge.
Zagzebski, L. 2003. Emotion and Moral Judgment. *Philosophy and Phenomenological Research* 66 (1): 104–124.

Chapter 4
Two Theories of Moral Cognition

Julia Haas

Abstract Moral cognition refers to the human capacity to experience and respond to situations of moral significance. Recently, philosophers and cognitive scientists have turned to reinforcement learning, a branch of machine learning, to develop formal, mathematical models of normative cognition. One prominent approach, proposed by Cushman (Curr Opin Behav Sci 3:58–62, 2015), suggests that moral cognition is underwritten by a habitual ('model-free') system, in conjunction with socially-learned moral rules. I argue that moral cognition instead depends on three or more decision-making systems, with interactions between the systems producing its characteristic sociological, psychological, and phenomenological features. Adopting such an approach allows us to not only better explain what is going on in everyday, 'successful' instances of moral judgment and action, but also to more reliably predict, and perhaps thereby counter, routine breakdowns in moral behavior.

Keywords Moral psychology · Moral cognition · Formal models · Reinforcement learning · Domain-generality

4.1 Introduction

Moral psychology investigates how we experience and respond to situations of moral significance (Doris and Moral Psychology Research Group 2010). Empirically-informed moral psychology proposes that the field be informed by relevant findings in the human sciences with an eye to advancing both descriptive and normative aims. The descriptive agenda involves understanding and predicting everyday moral experiences and actions, often described as 'moral cognition.' The normative agenda includes intervening on and ameliorating the mechanisms underlying moral cognition with the goal of enhancing human happiness, well-being, and welfare. Ideally, cooperation on both of these goals should lead to bi-directional progress (Doris et al. 2017).

J. Haas (✉)
Department of Philosophy, Rhodes College, Memphis, TN, USA
e-mail: julia.haas@anu.edu.au

G. S. Holtzman, E. Hildt (eds.), *Does Neuroscience Have Normative Implications?*, The International Library of Ethics, Law and Technology 22,
https://doi.org/10.1007/978-3-030-56134-5_4

Empirically-informed moral psychology has made progress on elucidating the components of moral cognition (Greene 2015). It is now widely accepted that affective experiences, and most notably emotions, have a profound influence on moral cognition (Nichols 2004; Haidt 2012; D'Arms and Jacobson 2014). It has also been emphasized that we take factors such as consequences and personal harm into consideration when making moral judgments, supporting the conclusion that moral cognition involves rule-based inference and reasoning more broadly construed (May 2018). Nonetheless, there is a growing sense that the field would benefit from a unified model of moral cognition.

Crockett (2016, p. 85) characterizes the current state of empirically-informed moral psychology in the following way:

> Imagine you want to bake a cake. The crucial first step is determining what ingredients are necessary for the cake—flour, sugar, eggs, milk, and so on. Next, you would need to know the amounts of each ingredient, and in what order to mix them. Similarly, past research in moral psychology has focused primarily on the critically important first step of identifying the key ingredients of moral judgments and decisions—norms, empathy, intentions, actions, outcomes, and so on. Now that many of these ingredients have been identified, future work can begin to develop formal mathematical models that describe how they are combined, in what amounts and in what temporal order, to produce moral judgments and decisions.

Philosophers and cognitive scientists have turned to theories from computer science and cognitive neuroscience with the objective of developing such formal mathematical models of normative cognition. Perhaps most prominently, philosophers and cognitive scientists have turned to an area in computational neuroscience known as *reinforcement learning,* which studies how agents learn through interactions with their environments, to try to understand the nature of moral cognition (Shenhav and Greene 2010; Crockett 2013, 2016; Cushman 2013, 2015; Crockett et al. 2017).[1]

Reinforcement learning research suggests that human beings depend on at least three computational algorithms and, by extension, on three semi-autonomous decision systems, for deliberation, choice, and action. The *hardwired* ('Pavlovian') system relies on automatic approach, withdrawal responses to appetitive and aversive stimuli, respectively (Macintosh 1983). The *habitual* ('model-free') system caches positive and negative state-action pairs. Finally, the *deliberative* ('model-based') system represents and selects from possible state-action pairs, often described in terms of a decision tree. Much of the present debate in this area centers on which of these systems best explain human moral judgments, choices, and behavior.

A prominent approach, proposed by Cushman (2015), suggests that moral cognition is underwritten by the habitual ('model-free') system, together with socially-learned moral rules. I argue, however, that although the model-free approach opens up a promising line of inquiry, it offers an unnecessarily narrow model of moral

[1] Reinforcement learning is an area in machine learning that aims to understand how agents learn to maximize rewards through interactions with their environments, i.e., how they learn through reinforcement, rather than through training on supervised data sets (Sutton and Barto 2018).

choice. That is, given that the reinforcement learning literature indicates that all three decision-mechanisms continually trade-off and interact to produce our everyday behaviors, there is no reason to accept Cushman's exclusive emphasis on the model-free decision system in the moral domain (Dayan and Niv 2008; Dayan and Daw 2008; Dayan and Berridge 2014; Huys et al. 2015; Garbusow et al. 2018).

Building on Cushman's model-free account, I defend a new, multi-system model of moral cognition. Specifically, I argue that moral cognition depends on three or more decision-making systems, with interactions between the systems producing its characteristic sociological, psychological, and phenomenological features. Adopting such an approach allows us to not only better explain what is going on in everyday, 'successful' instances of moral judgment and action, but also to more reliably predict, and perhaps thereby counter, routine breakdowns in moral behavior.

I begin by sketching a set of features of moral cognition in need of explanation. I then go on to introduce the basic principles of reinforcement learning, before turning, in the main body of the paper, to the application of these principles to competing explanations of moral cognition. I conclude by considering some of the implications of my view for the normative agenda of intervening on human moral decision-making and, by extension, potentially ameliorating human welfare.

4.2 Moral Cognition: What Needs Explaining?

Imagine a shopper named Barbara. Barbara is standing in the pasta aisle of her local grocery store, preparing to buy her usual brand of pasta. She then remembers that the company refused to feature a homosexual couple in its advertising.[2] Upon consideration, Barbara feels that it would 'just be wrong' to buy this brand and, seeing that a typically more expensive brand of pasta is on sale, she decides to buy this second type instead. Notably, Barbara remains unsure about which brand of pasta she will buy over the long term.

Barbara's pasta-buying experience is an example of *moral cognition*, or the capacity to experience and respond to situations of moral significance.[3] The experience raises a number of questions for empirically-informed theories of moral psychology. How do human beings make choices involving moral dimensions? Why do moral principles affect individuals differently at different times? And where does

[2] This example is loosely based on controversial remarks made by Guido Barilla, of Barilla pasta, in 2013. The company has since improved its record on LGBTQ rights (Somashekhar 2014).

[3] By cognition, I mean the general functioning of the mind, rather than the narrower mental processes associated with understanding and knowledge. Moral cognition here refers to our capacity for lay moral experience, including the experience of moral emotions such as anger or shame. In other branches of philosophy, moral cognition is sometimes referred as *moral perception*; however, this latter term has technical connotations that I do not want to import in the discussion here (for a review, see McGrath 2018).

the feeling that so often accompanies moral choices, namely that something is just 'right' or 'wrong,' come from?

However, one challenge facing the empirically-informed, moral psychological study of moral cognition is that it is a complex and multi-faceted capacity. No one feature seems to get at the essence of the human capacity for moral experience. Moreover, moral cognition is not a regular mechanism in the way that other targets of explanation are reliable: while we may occasionally feel compelled to do what is right, we just as frequently abandon our moral commitments to do what is pleasant or even expedient (and as we will see in this section, this punctuated compliance has both psychological and phenomenological components). How best to proceed, then, in our investigations of such a challenging subject of inquiry? And, perhaps even more difficult, how to choose between competing theoretical explanations of it?

In this section, I characterize moral cognition using a cluster approach. That is, I propose to draw on extant, broadly agreed-upon analyses of moral norms, judgments, and emotions to characterize moral cognition in terms of its core sociological, psychological, and phenomenological features. I further propose to use these features to assess and adjudicate between competing theoretical alternatives. Here, the working assumption is that the greater the number of features a given theory can account for, the better its standing vis a vis competing alternatives.

4.2.1 Sociological Features of Moral Cognition

Philosophers typically emphasize the role of moral norms in our human capacity for moral cognition. But from more empirically-informed perspectives, moral norms, and norms in general, are in fact most commonly studied from a sociological perspective, particularly using game-theoretic approaches. On these latter methodologies, moral norms are sometimes said to represent a special class of the 'social grammar' of social norms: moral norms are the products of human interaction, and they dictate what is and is not morally appropriate for members of a given group or society (Bicchieri 2006).

In the context of this sociological approach, moral norms are broadly considered to be ubiquitous and pervasive. There is evidence of moral norms in all past and present human societies. Further, Sripada and Stich (2005) suggest that norms follow reliable patterns of distribution, which is to say that norms cluster around general themes, such as prohibitions of killing and incest. At the same time, there is variability in how these normative themes become implemented into specific rules or 'maxims.' For example, what does and does not constitute acceptable grounds for a violent altercation can vary a great deal between different groups and communities. Typically, there also some exceptions to the norms held by a given group or society.

We can thus call *variable ubiquity* a sociological feature of moral cognition: moral cognition is reliably present among all human communities, although the contents involved in those moral cognition processes can vary to a relatively substantial degree.

4.2.2 Psychological Features of Moral Cognition

There is growing moral psychological interest in our capacity to assess, cognize, and respond to moral circumstances, complementing the substantial sociological and game-theoretic literature on moral cognition. In addition to the sociological feature of variable ubiquity, this body of literature highlights four psychological features of moral cognition. First, the developmental literature indicates that moral cognition exhibits a *reliable pattern of ontogenesis*, which is to say that typically developing human beings reliably develop the capacity to process and engage with the moral components of experience. For example, children as early as 3 years of age appear able to distinguish between non-moral and moral concerns (Smetana and Braeges 1990), to assess the role of intentions in assigning moral praise and blame (e.g., Nobes et al. 2009), and to protest when a third party is subject to moral transgressions (Rossano et al. 2011).

Second, moral cognition is widely thought to be *cognitively heterogenous*: it can be executed both quickly and intuitively, as when people automatically respond to an everyday moral circumstance, e.g., helping a person in immediate need. Alternately, moral cognition can be based on careful internal deliberation, as when people turn over their moral alternatives in their minds, particularly in facing a difficult moral choice.[4] This cognitive heterogeneity is often characterized in terms of dual processing accounts (Haidt 2012; Cushman et al. 2010; Campbell and Kumar 2012; Greene 2013; May 2018), where moral cognition depends on both a fast, automatic, and emotional system, as well as on a slow, demanding, and reasons-based system. One prominent, dual-process account further linked the automatic, emotional system with deontological, or 'duty-based' moral responses, and the slow, reasons-based system with utilitarian, or 'consequences-based' moral reasoning (Greene et al. 2001; Greene 2014). For our purposes, in an effort to avoid over-theorizing the phenomenon in advance, we can describe this psychological feature of moral cognition in much more open-ended terms, namely, by simply recognizing that the capacity for moral engagement appears to rely on more than one form of cognitive processing.

Third, moral psychologists frequently argue that moral cognition is reliably associated with certain types of *emotions*. For example, Prinz and Nichols (2010) note that moral cognition is regularly associated with self-blaming emotions such as guilt and shame, and other-blaming emotions such as anger. Along similar lines,

[4]This latter, deliberation-based form of moral cognition is different from a still third kind of moral capacity, namely, moral reasoning or debate – the sort of reasoning on display in the pages of philosophy journals, where we aim to develop general and intermediate moral principles, defend and apply them, and so on.

Kelly (2011) argues that the emotion of disgust plays a central role in moral cognition. Notably, one does not have to be a *sentimentalist* about moral cognition, that is, one does not have to hold that emotions are *constitutive* of our moral capacities, in order to recognize that there is a reliable association between moral responses and certain types of emotions (May 2018).

Fourth, moral cognition can be characterized by the aforementioned *irregularity* that we so often associate with moral behavior: that is, we typically comply with our moral principles and values, but there are not infrequent exceptions to this rule (Doris 2002). In other words, as individuals, we take moral principles and rules to be compelling, although even a brief consideration reveals that, apart from social and punitive repercussions, nothing really obliged us to do the right thing and that, indeed, we sometimes fail to do what is good. This last psychological feature of moral cognition represents a particular challenge for mechanistic explanations since, in principle, mechanistic explanations aim to account for patterns and *regularities* in a given phenomenon.

4.2.3 Phenomenological Features of Moral Cognition

Finally, and in some respects most intriguingly, it may be said that moral cognition is associated with a distinctive phenomenological 'feel.' To be moved by what is right – or, conversely, to be moved to avoid doing what is wrong – is typically accompanied by a certain feeling of motivation or force or, as Cushman calls it, an accompanying feeling of '*moral constraint*' (2015). There is a special sense in which moral considerations move us in ways that other types of considerations do not. The feeling of moral constraint is thus a core feature of moral cognition that must be accounted for, and either be assimilated with or distinguished from other types of motivation.

The next section provides a brief overview of these reinforcement learning approaches, before turning to its application to moral cognition.

4.3 Reinforcement Learning

4.3.1 Basic Features

The reinforcement learning research program broadly refers to normative, computational approaches to understanding how agents learn to maximize rewards through interactions with their environments. Early theories proposed that learning only occurs when a single event violates an agent's expectations (Rescorla and Wagner 1972; see also Schultz et al. 1997). A subsequent theory (Sutton and Barto 2018) developed this foregoing principle to analyze how agents learn to maximize their rewards over time. Here, we focus on Sutton and Barto's reinforcement learning framework, together with subsequent developments, to show how principles of the theory are used to analyze choice and action.

Reinforcement learning models generally consist of four main components: a set of states, a set of actions, rules governing the transitions between states, and rules that determine the immediate reward associated with a given transition and corresponding state. Notably, in reinforcement learning environments, an agent does not know how the transitions work, or what the rewards will be in each state. Instead, she must learn from experience and try to predict the values of transitions and rewards that she has not yet made. To do so, an agent can deploy one or more computational strategies to try and best estimate the value of different courses of action, and ultimately to arrive at a policy that will enable her to collect as much reward as possible.

The current consensus in the literature suggests that decision-makers approaching optimality – including human beings – rely on not one but at least three dissociable decision-making systems to make choices in their environment (for two excellent introductions, see Montague 2006; Redish 2013; for more specialized reviews, see Dayan and Abbott 2001; Dayan and Niv 2008; Rangel et al. 2008).

4.3.2 The Three Systems

We can now discuss each of the three systems in detail. Behaviors issued by the first of the three systems, known as the *Pavlovian* system, are characterized by automatic approach and withdrawal responses to appetitive and aversive stimuli, respectively (Macintosh 1983). For example, we 'naturally' *approach* pieces of tasty food, but *avoid* scary snakes or spiders. These general types of behaviors are thought to be underwritten by the evolutionarily old, 'hardwired' Pavlovian systems. Notably, though Pavlovian responses are appropriate in natural environments, since it is broadly beneficial to approach rewards and avoid punishments, these responses can also lack flexibility, which can result in detrimental outcomes (Huys et al. 2012).

Due to this characteristic behavioral rigidity, researchers identify Pavlovian responses by persistence even in those cases where this ongoing, 'stubborn' response is detrimental. For example, David and Harriet Williams demonstrated that pigeons continue to peck at a key associated with food, even when doing so resulted in losing the reward (Williams and Williams 1969; see also Brown and Jenkins 1968). Similar response patterns have been found with other animals, including in human beings (Sheffield 1965; Hershberger 1986; Bouton 2006; Redish 2013).

A second, *model-based* system explicitly represents possible choices and determines the sequence of actions that maximizes value (Dayan 2011; Daw and O'Doherty 2013). This procedure is typically represented by a *decision tree*. Each node in the tree represents a possible choice. The model-based system searches through the decision tree to find the branch with the highest total value. Hence its alternative name: tree search. For example, a chess player may represent three upcoming moves in a game of chess, with each possible move further branching into a wide range of subsequent moves. To win, the player then tries to represent and choose the best possible sequence of moves overall. Of course, representing all of one's future moves in chess is famously challenging, because the options quickly branch beyond any reasonable ability to explicitly represent them. Consequently, whereas Pavlovian responses are automatic and inflexible, the model-based system responds flexibly to new situations but struggles if there are more than a few options to consider.

A third, *model-free* system does not explicitly represent future alternatives, but caches positive and negative experiences, and assigns values to actions based on their previous outcomes (Dayan 2011; Daw and O'Doherty 2013). To cache experiences, the model-free system employs a feedback signal, which revises the system's estimates about the environment. Specifically, the model-free system is thought to be described by the so-called temporal difference (TD) learning algorithm, which enables an agent to weigh expectations against actual rewards and use the comparisons to optimize actions over time. If the reward is larger than expected, the signal indicates that things have gone 'better than expected'; if the reward turns out to be smaller than anticipated, the signal registers 'worse than expected'; if the anticipated and experienced rewards are the same, the signal records 'no change.' On this approach, 'good' state-action pairs are those that have produced rewarding outcomes in the past, and so should be repeated. Conversely, 'bad' state-action pairs have produced punishments in the past, and so should be avoided.[5]

In virtue of its caching-based approach, model-free learning exhibits the opposite computational strengths and weaknesses of those of model-based learning. As its name suggests, the model-free system does not apply the modeling stages of model-based learning. Rather, it interacts with the environment and updates its estimation of the best course of action over time. One important byproduct of this approach is that the model-free system is not immediately sensitive to unexpected change. Due to the 'caching' nature of the algorithm, it instead takes several low-reward experiences to lower the overall high values that had previously been attributed to a state. This 'recalcitrant' response is taken as a signature feature of model-free learning – a feature that will become important in our discussions of moral cognition. On the other hand, over time, the system is efficient, because, unlike the model-based system, it does not need to explicitly represent all of the action possibilities in every given decision.

[5] I owe this formulation to Crockett (2013).

4.3.3 *Linking the Normative Models to the Brain*

The greatest discovery linking normative reinforcement models to actual biological processes was Wolfram Schultz, Peter Dayan, and Read Montague's discovery that the TD learning signal accurately predicts the firing of dopamine neurons in relation to unexpected rewards (Schultz et al. 1997; though see Langdon et al. 2018; Gardner et al. 2018). Montague describes the discovery as follows. Contrary to the then-accepted view that dopamine reflects experiences of pure, immediate reward, "Schultz noticed that dopamine neurons change their activity when 'important' events happened, like a juice squirt, or the appearance of food, or even a sound in the laboratory that predicted that food or drink was about to be delivered" (Montague 2006, p. 108). This led him to the possibility that dopamine could correspond to a kind of prediction error. When he looked at Schultz's data, Dayan immediately "recognized a striking resemblance between dopamine neuron activity and error signals used in abstract reinforcement learning algorithms… it was an amazing match. The model showed that Schultz had discovered one of the central critic systems in the mammalian brain, and one that encoded its criticism in the delivery of dopamine" (Montague 2006, p. 109). Over the subsequent decade, findings have shown that all three systems have been found to play a prominent predictive role in decision-making and action, including in human decision-making and action (for helpful overviews, see Dayan 2014; Dayan and Berridge 2014; Daw and Tobler 2014; Moutoussis et al. 2018).

One primary, experimental way of distinguishing between the model-free and model-based decision systems is known as *post-training reward devaluation* (henceforth 'reward devaluation') (Dickinson and Balleine 2002; see also Tricomi et al. 2009). In reward devaluation, an agent is trained to perform a certain action in exchange for a reward, e.g., pushing a button in exchange for a treat. The period of this training is either short, such that the agent continues to rely on her forward-looking, model-based decision system; or it is somewhat longer, such that the agent comes to rely on her caching-based, model-free system, i.e., that she can habituate the relevant activity. In experiments using this paradigm, the reward is then devalued, e.g., in the case of rats, by pairing it with a nausea-inducing chemical or, in the case of human participants, simply by overfeeding it to them, to the point where the reward is no longer pleasant to consume, and the agent's willingness to perform the task is then analyzed. In those cases where the training period was sufficiently *short*, the agent can respond 'appropriately' in the sense of no longer performing the action that generates more of the now-unpleasant reward. But in those cases where the training period was *longer*, and thus led to dependence on the model-free system to perform the instrumental action, the agent responds 'inappropriately,' that is, she continues to perform the task, even though the reward is no longer appealing her.

Emerging lesion studies further complement these findings. For instance, the orbitofrontal cortex has been shown to a particular role in representing model-based values (Valentin et al. 2007). Lesions to the dorsolateral striatum and nigrostriatal dopamine systems prevent an animal from developing model-free learning (Yin et al. 2004; Faure et al. 2005). And while relatively little is known about the structures underlying potential arbitration between the two valuation systems, some evidence is beginning to emerge. In particular, two candidates for arbitration are being proposed in the shape of the infralimbic cortex and the anterior cingulate cortex. Lesions to the infralimbic cortex reintroduce model-based learning from what had previously relied on the model-free controllers, and studies have shown that the anterior cingulate cortex is involved in monitoring and resolving response errors and conflict (Daw et al. 2005, p. 1708).

Admittedly, having multiple systems competing to evaluate a single task might seem at best inefficient and at worst disadvantageous. However, simulations suggest that optimal agents in fact employ several complementary controllers (Dayan 2011; Daw et al. 2005). As Daw and colleagues note, "The difference in the accuracy profiles of different reinforcement learning methods both justifies the plurality of control and underpins arbitration. To make the best decisions, the brain should rely on a [system] of each class in circumstances in which predictions tend to be most accurate" (Daw et al. 2005, p. 1704).

One line of inquiry has analyzed how a possible *principle of arbitration* can dictate the circumstances under which a given system should govern a specific choice problem (Daw et al. 2005; Lee et al. 2014). A prominent approach proposes that interactions between the systems are governed by the following 'accuracy-based principle of arbitration' (henceforth, 'PA') (adapted from Daw et al. 2005):

PA: Following partial evaluation, that system with the highest accuracy profile, i.e., that system most likely to provide an accurate prediction of expected value, relative to the decision problem at hand, directs the corresponding assessment of value.

In this view, each of the multiple systems partially evaluates the action alternatives. Simultaneously, each system generates an estimate of how accurate its prediction is relative to the decision problem at hand. These estimates, or accuracy profiles, are then compared, and the system with the highest accuracy profile is selected to direct the corresponding valuation task. PA thus directs valuation based on a system's accuracy profile, rather than on its prediction of value (Deneve and Pouget 2004; Daw et al. 2005; Lee et al. 2014). For example, the Pavlovian system typically coordinates choice in familiar, complex settings, because it typically has a higher accuracy profile in those decision problems, even if it predicts a lower overall value than do either its model-based or model-free counterparts.

A related line of inquiry suggests that all three decision-mechanisms continually trade off and interact to produce even a single, everyday behavior (Dayan and Niv 2008; Dayan and Daw 2008). For example, it is thought that the model-based system may be used to train up its model-free counterpart, while some evidence suggests that both the Pavlovian system is systematically used to 'prune' branches in model-based decision-trees in order to overcome the latter's signature computational complexity (Huys et al. 2012).

Taking both of the lines of inquiry together, a picture emerges on which all three systems continually interact and trade-off to produce our everyday choices and actions. In what follows, we will see that this rich network of interactions may help explain the multi-faceted and sometimes perplexing nature of everyday moral cognition. But first, let's look at Cushman's influential (Cushman 2015) proposal regarding reinforcement learning, model-free learning, and moral cognition.

4.4 Cushman's Model-Free Approach to Moral Cognition

4.4.1 The Core Thesis of Cushman's View

Fiery Cushman (2015) considers the mechanisms underpinning human moral cognition with a special emphasis on moral constraint. Cushman suggests that the question of moral constraint is of particular interest insofar as the feelings associated with moral cognition seem to compel us in ways that other kinds of considerations do not. Cushman argues that these feelings and the corresponding behaviors are underwritten by the model-free decision system.

Cushman makes his case by drawing on the signature feature of model-free learning. Recall from the previous section that model-free learning does not represent future alternatives, but instead caches positive and negative experiences and predicts future value accordingly. Recall also that this caching procedure means that the system is not immediately sensitive to unexpected changes. The central premise in Cushman's argument is that moral behaviors exhibit precisely this signature feature of model-free learning, namely, this lack of sensitivity to changes in reward value. Hence, moral actions, including altruistic actions, are broadly thought to be supported by automatic, i.e., model-free responses (Rand et al. 2014). Specifically, Cushman argues that when the brain produces the affective component of moral cognition and constraint, such as a feeling of obligation, "we assign intrinsic value to the action, rule or norm in question," such that it is difficult to act otherwise than in accordance with it (Cushman 2015, p. 60–61). For example, Cushman argues, American tourists frequently continue to tip in restaurants abroad, even when there is no relevant norm dictating that they should do so (Cushman 2015, p. 59). Related examples include apologizing where an apology is not really necessary, or quickly helping a person in need without first stopping to think about what the right thing to do would be.

Cushman illustrates his thesis by drawing on the model-free explanation to explain participants' inconsistent responses to the trolley problem. Famously, in 'switch' versions of the trolley problem, where the cause of a hypothetical harm is relatively indirect, participants are more likely to support the killing of a single individual in order to save five others. Inconsistently, however, participants are more likely to find it difficult to endorse the harm of one agent in 'footbridge' versions of the problem, where the harm involves pushing another individual and so is more 'hands on' (Greene et al. 2001).

Since a purely numerical assessment favors the saving of five people rather than one in *both* cases, Cushman reasons, people's tendency to resist harming the single agent in the footbridge version may be "the consequence of," and so explained by, "negative value assigned intrinsically to an action: direct, physical harm" (Cushman 2015, p. 59). That is, Cushman suggests, participants' responses to the footbridge version of the dilemma may be underwritten by the model-free decision-system. Since directly harming others has reliably elicited punishments in the past, this option represents a bad state-action pair, on the model-free view, and leads people to reject it as an appropriate course of action.

4.4.2 *Assessing Cushman's View with Respect to the Features of Moral Cognition*

How does Cushman's model-free view fare with respect to the sociological, psychological, and phenomenological features of moral cognition outlined above?

On my view, Cushman's account has both significant explanatory advantages and important explanatory disadvantages. The proposal's central advantages stem from its appeal to the domain-general, model-free decision system. A domain-general decision system is one which can be applied to all kinds of decisions, e.g., ranging from economic choices between goods to choices about how best to act. Explaining moral cognition in terms of such a domain-general system is advantageous because it avoids positing a dedicated system or mechanism for explaining our moral experiences. As Greene (2015, p. 40) observes, Cushman's account joins other accounts in suggesting that "what we call 'moral cognition,' is just the brain's general-purpose cognitive machinery— machinery designed to learn from experience, represent value and motivate its pursuit. represent mental states and traits, imagine distal events, reason, and resist impulses – applied to problems that we, for high- level functional reasons, identify as 'moral.'?"

Cushman's domain-general approach also helps explain moral cognition's characteristic ubiquity, ontogenesis, and motivational phenomenology. We can start with the ubiquity of moral cognition. As noted in Sect. 4.2, moral cognition is thought to be a ubiquitous feature of human experience. But ubiquity is precisely what we should expect if moral cognition depends on a much older, more general cognition system such as the decision-making system. Moreover, if this approach to moral cognitive ubiquity is right, then we should also expect that *breakdowns* in moral cognition are associated with related breakdowns in the more general capacity, namely, in this case, in the more general capacity to make decisions.

This is indeed what we find. Empirically-informed moral psychologists commonly cite psychopaths as paradigmatic examples of individuals who fail to exhibit typical moral cognitive capacities, because they seem in many in ways incapable of making typical moral judgments and decisions (Blair 1995; Nichols 2002; Blair 2017; though see Jalava and Griffiths 2017). But psychopaths exhibit corresponding, general deficits in decision-making, for example, in their performance on decision-making tasks such as the Iowa Gambling Task (Mahmut et al. 2008). In this regard, Cushman's proposal that moral cognition is supported by the model-free decision system falls squarely within a growing body of literature regarding the relationship between moral and general decision-making (Shenhav and Greene 2010). Of course, this does not yet suggest that moral cognition *only* involves domain-general decision systems – only that these systems at least play a role.

An analogous line of reasoning can be presented regarding the reliable ontogenesis of moral cognition. If moral cognition is underwritten by the model-free system, and so depends on this more general decision-making capacity, then we should expect children's moral capacities to 'come online' at around the same time that they begin to exhibit adaptive, automatic decision-making abilities, e.g., on modified versions of the Iowa Gambling task. And indeed, although the data is less clear here than in the case of psychopaths, findings do suggest that children do begin to make advantageous choices on decision tasks at around the same time that they begin to recognize the moral features of their environment (Crone and van der Molen 2004; Killen and Smetana 2015).

Similarly, finally, the domain-general nature of the model-free decision system explains the characteristic phenomenological experience of moral constraint. It is generally thought that reinforcement learning, including the model-free system, explains how choices are transformed into corresponding actions, i.e., how choices motivate action (Glimcher and Fehr 2013). Correspondingly, if we think of moral cognition as supported by this more general system, then we can equally expect it to explain the phenomenology associated with specifically *moral* motivation. Indeed, one of the central implications of Cushman's view is that moral motivation may operate in roughly the same way it does in other decisions, albeit perhaps more forcefully (Cushman 2015, p. 60).

On the other hand, the view's exclusive emphasis on the model-free decision system limits its overall explanatory power. That is, the model-free view struggles to explain moral cognition's cognitive heterogeneity, punctuated compliance with moral norms, and reliable association with moral emotions. Given that Cushman's view exclusively focuses on model-*free* learning, it struggles to explain how we might explicitly represent or deliberate about a given moral dilemma. Similarly, if we think of the model-free system as supporting something akin to habituation, it is hard to reconcile this computational approach with the observation that we routinely – but irregularly – fail to act in ways that correspond with our moral norms and judgments. And the model-free view is simply *neutral* when it comes to presence of moral emotions in moral cognition. That is, the model-free view is neither at odds with this feature of moral cognition, nor does it have any specific resources to explain why these emotions are so reliably present.

4.4.3 Two Objections and a Proposal

Despite these explanatory limitations, the model-free view of moral cognition represents a highly promising alternative to extant, highly influential dual-system views of moral judgments and decision-making (Driver 2016). Nonetheless, Cushman's account faces two more general objections.

First, it is important to recognize that all three of the view's foregoing explanatory successes rely on the domain-general nature of model-free learning, rather than on the specific features of the model-free system itself. This is not problematic in and of itself, but it is worth recognizing that either or both of the other two decision systems share this explanatory power. That is, the model-free explanation of ubiquity, ontogenesis, and moral constraint is more generally a *reinforcement learning-based* explanation, and so is not exclusively available, in explanatory terms, to the narrower view Cushman puts forward.

Second, I argue that Cushman's proposal is simply unnecessarily restricted in its use of the reinforcement learning toolkit. That is, given the growing body of evidence which suggests that all three decision-mechanisms continually trade off and interact to produce our everyday behaviors, Cushman's exclusive emphasis on model-free learning is problematic (Dayan and Niv 2008; Dayan and Daw 2008; Dayan and Berridge 2014; Huys et al. 2015). Correspondingly, it seems implausible that normative actions alone are exclusively underwritten by the model-free mechanism, as Cushman's view suggests.

We can illustrate this latter point by applying it to Cushman's assessment of the trolley problem. In particular, more evidence would be needed to establish that it is exclusively the model-free system that is involved in participants' responses to the footbridge version of the trolley problem. This is because, if we recall from above, *two* systems are known to respond sub-optimally in the face of a change in rewards: both the model-free *and* the Pavlovian systems respond in this way (see Crockett 2013). The difference is that while the Pavlovian system is *permanently* insensitive to reward devaluation, the model-free system slowly adapts to reward devaluation over time (albeit much more gradually than its model-based counterpart). Within the reinforcement learning framework, there are thus *two* plausible interpretations of peoples' responses to the physical harm involved in the footbridge dilemma. On Cushman's hypothesis, the participants' aversion to harm is rooted in the model-free valuation of a habitually negative state-action pair. Alternatively, the participants' aversion may be underwritten by a much more deeply rooted, Pavlovian response to overwhelmingly aversive stimuli (Huys et al. 2012).

One way to disentangle these competing theories would be to conduct a targeted version of the trolley experiment, designed to test for the role of either the model-free or Pavlovian learning systems in people's responses to the problem. Using a head-mounted display device, the participants would be placed in a virtual reality environment (VRE) that imitates classic trolley problem scenarios (VRE-based methodology adapted from Navarrete et al. 2012). Participants would be told they have two alternatives to choose from in each simulation: to stop an oncoming train, or to do nothing.

The participants would be divided into two conditions. In the *switch condition,* a participant would appear to be standing near a railway track, with five railway employees shown to be repairing the tracks further on down the main line. There would also be a switch next to the participant's right hand (physically represented by a joystick in the lab). Left as is, the switch would allow a virtual train to continue straight on the main track, killing the five employees. If pulled, the switch would divert the train onto a sidetrack and kill a single individual. In the *footbridge condition*, the participant would appear to be standing on a footbridge spanning a railway track, while the five railway employees will be shown repairing the tracks below. In addition, a virtual pedestrian would be standing on the footbridge directly in front of the participant. Pushing the fellow pedestrian onto the track, physically represented by a soft mannequin in the lab, would cause the train to stop before hitting the five employees. Not pushing the pedestrian onto the track would result in the train continuing down the main track and killing the five individuals. In both conditions, sounds of distress coming from either the one or five agents would become audible, depending on the participant's choice.

The key feature of the experiment would consist of repetition. Recall that the Pavlovian and model-free systems are both 'slow to update.' But this is not quite accurate. The model-free system is slow to update; the hardwired, Pavlovian system *never updates at all*. Hence, the participants would be asked to experience multiple versions of both conditions. If Cushman's hypothesis is correct, the participants' responses to the footbridge condition would gradually update to reflect more 'utilitarian' responses. If the alternative hypothesis is correct, the participants' responses to the footbridge condition would remain consistent in their refusal to push the imaginary individual off of the footbridge to save the five individuals on the tracks.

In the next section of the paper, I propose a view that accommodates both hypotheses.

4.5 A Multi-system View of Moral Cognition

At the outset of this paper, I proposed to use Cushman's model-free account as a jumping-off point for a novel, multi-system, reinforcement-learning-based model of moral cognition. We are now in a position to take this step. In contrast to Cushman (2015), I argue that the Pavlovian, model-based, and model-free systems are all implicated in normative cognition.

4.5.1 The Multi-system View

Many of the pieces of the multi-system view are already in place. It stipulates that moral cognition is underwritten by three domain-general, decision systems: namely, the Pavlovian, model-based, and model-free decision systems. As we will see in what follows, many of the core features of moral cognition reflect the signature profiles of these three decision systems.

In addition, these three systems interact to produce additional types of decision-making responses. Interactions between these systems are governed by the principle of arbitration characterized above, namely, PA. These interactions can take two general forms: first, PA may be used to a 'select' a 'winning' decision system, which goes on to make a corresponding, valuation-based decision; or, alternately, second, PA can dictate that two or more systems trade-off on a single decision, as when the Pavlovian system prunes the model-based decision tree in order to simplify the choice matrix.

Importantly, by taking all three systems into consideration, together with the ongoing interactions between them, the multi-system model can explain the six core features of moral cognition.

4.5.2 Assessing the View

In addition to Cushman's emphasis on the model-free system, the forward-looking nature of model-based learning enables us to explain the deliberative components of moral cognition and, by extension, its cognitive heterogeneity. Moral reasoning often involves such capacities as setting goals, deliberating about what it would be best to do, and reconciling competing moral demands. These are the signature features of model-based learning: an agent identifies a specific goal, evaluates the different available means of achieving it, and pursues what she has identified as the best alternative. At the same time, the computationally taxing nature of model-based learning may help explain the use of explicit deliberation mainly in relatively novel or high-stakes moral situations, rather than in more familiar or low-stakes settings, as in Cushman's (2015) example of tipping in a restaurant, discussed above. The multi-system account thus provides a comprehensive approach to understanding the heterogeneous nature of moral cognition.

For their part, given their rapid, hardwired, and recalcitrant nature, Pavlovian responses most obviously figure in moral cognition as emotional responses. A wide range of theories highlights the importance of various emotions in normative behavior, emotions including guilt, shame, and anger (for a review, see Prinz and Nichols 2010). But emotional responses are thought to themselves be a basic type of Pavlovian action, triggering automatic changes in heart rate, facial expression, and so on (Redish 2013).

From a slightly different perspective, recall that Pavlovian responses are characterized by the fact that they persist well beyond optimality, as when chickens continue to peck at a dispenser even when it means not receiving their food. But this is precisely the kind of response we see in certain moral social interactions. For instance, experiments based on the ultimatum game, which ask participants to accept or reject monetary offers from a confederate researcher, show that participants frequently reject offers that they perceive to be unfair, even if doing so prevents them from receiving any money at all (Van't Wout et al. 2006). These decisions can thus be seen as a form of the recalcitrant or suboptimal decision characterized by Pavlovian responding.

More controversially, Pavlovian responding may also play a role in the phenomenon known as *moral dumbfounding*. Moral dumbfounding can be defined as the "stubborn and puzzled maintenance of a judgment without supporting reasons" (Haidt et al. 2000, p. 1). For example, in a classic example of the phenomenon, subjects continue to believe that sex between siblings is wrong, even when they are told that all potential sources of harm have been accounted for. Insofar as these subjects maintain their judgments even after it no longer makes sense to do so, however, their responses exhibit the characteristic feature of Pavlovian decision-making.

Contra Cushman (2015), the properties of all three decision systems thus play a role in explaining the core features of moral cognition. However, I argue, finally, that it is the *interactions* between systems that explain the irregular or punctuated nature of compliance with moral judgments and beliefs. We often think of moral cognition as operating smoothly until 'something goes wrong.' That is, the multi-system view suggests that these intermittent lapses may instead be the products of the systems' regular workings (Haas 2018). The multiple decision systems operate in parallel, trading off to optimize decision-making. But in turn, these tradeoffs may occasionally result in responses that do not accord with our deliberative judgments of what it is best to do, or even with what, based on our habituated, model-free systems, we *typically* do. These tradeoffs can occur when the model-free system overrides its model-based counterpart or, alternately, when the Pavlovian system prunes the decision tree in order to simplify the decision process. As a result, we most often feel compelled by our moral principles and judgments, but also not infrequently fail to act in accordance with them.

4.6 Conclusion

Moral cognition is implicated in discussions of normative considerations such as moral dilemmas (Nichols and Mallon 2006), moral responsibility (Nahmias et al. 2014), moral luck (Cushman 2008), and the nature of the moral self (Strohminger and Nichols 2014). Nonetheless, it remains difficult to understand how best to effectively *intervene* on moral cognition and, by extension, on these explicitly normative issues.

In my view, the multi-system model offers a promising way of investigating possible interventions in moral cognition. This is because there is a close connection between understanding a mechanism and knowing how to intervene in it. As suggested by Crockett's cake analogy (Crockett 2016, p. 85), computational approaches to moral cognition are making important headway in terms of providing such 'recipes' – or mechanistic accounts – of how variously theorized 'ingredients' or components of moral cognition interact to produce our everyday human moral experiences. Such computational approaches can make increasingly accurate predictions about patterns of *breakdowns* in moral cognition, such as the relationship between the neurotransmitter serotonin and judgments about harm aversion (Crockett et al. 2010). By extension, computational approaches to moral cognition can offer important clues to those who would leverage these mechanisms with the goal of intervening on and improving human moral decision-making.

The multi-system model of moral decision-making extends this computational and mechanistic approach to moral cognition. In particular, the multi-system approach helps explain how our moral decision systems interact, and so enables us to recognize patterns of suboptimal decision-making that are produced as a consequence of these interactions. As I have argued elsewhere, in this sense, our decision systems and their interactions can help us understand the "fault lines" of our decision-making and, specifically, of our moral decision-making (Haas 2018). By extension, the multi-system model thus provides the preliminary foundations for a comprehensive set of principles to guide our moral choices and actions – and potentially begins to empower us to counteract such suboptimal responses if and when we sit fit to do so.

References

Bicchieri, C. 2006. *The Grammar of Society*. Cambridge: Cambridge University Press.

Blair, R.J.R. 1995. A Cognitive Developmental Approach to Morality: Investigating the Psychopath. *Cognition* 57 (1): 1–29.

———. 2017. Emotion-Based Learning Systems and the Development of Morality. *Cognition* 167: 38–45.

Bouton, M.E. 2006. *Learning and Behavior: A Contemporary Synthesis*. Sunderland: Sinauer.

Brown, P.L., and H.M. Jenkins. 1968. Auto-Shaping of the Pigeon's Key-Peck. *Journal of the Experimental Analysis of Behavior* 11 (1): 1–8.

Campbell, R., and V. Kumar. 2012. Moral Reasoning on the Ground. *Ethics* 122 (2): 273–312.

Crockett, M.J. 2013. Models of Morality. *Trends in Cognitive Sciences* 17 (8): 363–366.

———. 2016. How Formal Models Can Illuminate Mechanisms of Moral Judgment and Decision Making. *Current Directions in Psychological Science* 25 (2): 85–90.

Crockett, M.J., L. Clark, M.D. Hauser, and T.W. Robbins. 2010. Serotonin Selectively Influences Moral Judgment and Behavior Through Effects on Harm Aversion. *Proceedings of the National Academy of Sciences* 107 (40): 17433–17438.

Crockett, M.J., J.Z. Siegel, Z. Kurth-Nelson, P. Dayan, and R.J. Dolan. 2017. Moral Transgressions Corrupt Neural Representations of Value. *Nature Neuroscience* 20 (6): 879.

Crone, E.A., and M.W. van der Molen. 2004. Developmental Changes in Real Life Decision Making: Performance on a Gambling task Previously Shown to Depend on the Ventromedial Prefrontal Cortex. *Developmental Neuropsychology* 25 (3): 251–279.

Cushman, F. 2008. Crime and Punishment: Distinguishing the Roles of Causal and Intentional Analyses in Moral Judgment. *Cognition* 108 (2): 353–380.

———. 2013. Action, Outcome, and Value: A Dual-System Framework for Morality. *Personality and Social Psychology Review* 17 (3): 273–292.

———. 2015. From Moral Concern to Moral Constraint. *Current Opinion in Behavioral Sciences* 3: 58–62.

Cushman, F., L. Young., and J. Greene. 2010. Multi-system Moral Psychology. In *The Moral Psychology Handbook*, ed. J. M. Doris and The Moral Psychology Research Group, 47–71. New York: Oxford University Press.

D'Arms, J., and D. Jacobson. 2014. Sentimentalism and Scientism. In *Moral Psychology and Human Agency*, ed. J. D'Arms and D. Jacobson. New York: Oxford University Press.

Daw, N.D., and J.P. O'Doherty. 2013. Multiple Systems for Value Learning. In *Neuroeconomics: Decision Making, and the Brain*, 393–410. San Diego, CA: Elsevier.

———, N.D., and P.N. Tobler. 2014. Value Learning Through Reinforcement: The Basics of Dopamine and Reinforcement Learning. In *Neuroeconomics*, 2nd ed., 283–298. San Diego, CA: Elsevier.

———, N.D., Y. Niv, and P. Dayan. 2005. Uncertainty-Based Competition Between Prefrontal and Dorsolateral Striatal Systems for Behavioral Control. *Nature Neuroscience* 8 (12): 1704.

Dayan, P. 2011. Interactions Between Model-Free and Model-Based Reinforcement Learning,' Seminar Series from the Machine Learning Research Group. University of Sheffield, Sheffield. Lecture recording. http://ml.dcs.shef.ac.uk/. Accessed Oct 2018.

———. 2014. Rationalizable Irrationalities of Choice. *Topics in Cognitive Science* 6 (2): 204–228.

Dayan, P., and L.F. Abbott. 2001. *Theoretical Neuroscience*. Cambridge, MA: MIT Press.

Dayan, P., and K.C. Berridge. 2014. Model-Based and Model-Free Pavlovian Reward Learning: Revaluation, Revision, and Revelation. *Cognitive, Affective, & Behavioral Neuroscience* 14 (2): 473–492.

Dayan, P., and N.D. Daw. 2008. Decision Theory, Reinforcement Learning, and the Brain. *Cognitive, Affective, & Behavioral Neuroscience* 8 (4): 429–453.

Dayan, P., and Y. Niv. 2008. Reinforcement Learning: The Good, the Bad and the Ugly. *Current Opinion in Neurobiology* 18 (2): 185–196.

Deneve, S., and A. Pouget. 2004. Bayesian Multisensory Integration and Cross-Modal Spatial Links. *Journal of Physiology-Paris* 98 (1–3): 249–258.

Dickinson, A., and B. Balleine. 2002. The Role of Learning in Motivation. In *Learning, Motivation and Emotion*, ed. C.R. Gallistel, 497–533. New York: Wiley.

Doris, J.M. 2002. *Lack of Character: Personality and Moral Behavior*. New York: Cambridge University Press.

Doris, J.M., and Moral Psychology Research Group. 2010. *The Moral Psychology Handbook*. Oxford: OUP.

Doris, J., S. Stich, J. Phillips., & L. Walmsley. 2017. Moral Psychology: Empirical Approaches. In *The Stanford Encyclopedia of Philosophy,* Winter 2017 Edition. Edward N. Zalta (Ed.) Retrieved from: https://plato.stanford.edu/archives/win2017/entries/moral-psych-emp/

Driver, J. 2016. The Limits of the Dual-Process View. In *Moral Brains: The Neuroscience of Morality*, ed. S.M. Liao, 150–158. Oxford: Oxford University Press.

Faure, A., U. Haberland, F. Condé, and N. El Massioui. 2005. Lesion to the Nigrostriatal Dopamine System Disrupts Stimulus-Response Habit Formation. *Journal of Neuroscience* 25 (11): 2771–2780.

Garbusow, M., C. Sommer, S. Nebe, M. Sebold, S. Kuitunen-Paul, H.U. Wittchen, et al. 2018. Multi-Level Evidence of General Pavlovian-to-Instrumental Transfer in Alcohol Use Disorder. In *Alcoholism-Clinical and Experimental Research*, vol. 42, 128a. Hoboken: Wiley.

Gardner, M.P., G. Schoenbaum, and S.J. Gershman. 2018. Rethinking Dopamine as Generalized Prediction Error. *BioRxiv*: 239731.

Glimcher, P.W., and E. Fehr, eds. 2013. *Neuroeconomics: Decision-Making and the Brain*. 2nd ed. Waltham: Elsevier.
Greene, J. 2013. *Moral Tribes*. New York: Penguin Press.
———. 2014. Beyond Point-and-Shoot Morality: Why Cognitive (Neuro)Science Matters for Ethics. *Ethics* 124 (4): 695–726.
Greene, J.D. 2015. The Rise of Moral Cognition. *Cognition* 135: 39–42.
Greene, J.D., R.B. Sommerville, L.E. Nystrom, J.M. Darley, and J.D. Cohen. 2001. An fMRI Investigation of Emotional Engagement in Moral Judgment. *Science* 293 (5537): 2105–2108.
Haas, J. 2018. An Empirical Solution to the Puzzle of Weakness of Will. *Synthese* 12: 1–21.
Haidt, J. 2012. *The Righteous Mind*. New York: Pantheon.
Haidt, J., F. Bjorklund., & S. Murphy. 2000. *Moral Dumbfounding: When Intuition Finds No Reason*. Unpublished manuscript, University of Virginia.
Hershberger, W.A. 1986. An Approach Through the Looking-Glass. *Animal Learning & Behavior* 14 (4): 443–451.
Huys, Q.J., N. Eshel, E. O'Nions, L. Sheridan, P. Dayan, and J.P. Roiser. 2012. Bonsai Trees in Your Head: How the Pavlovian System Sculpts Goal-Directed Choices by Pruning Decision Trees. *PLoS Computational Biology* 8 (3): e1002410.
Huys, Q.J., N. Lally, P. Faulkner, N. Eshel, E. Seifritz, S.J. Gershman, et al. 2015. Interplay of Approximate Planning Strategies. *Proceedings of the National Academy of Sciences of the United States of America* 112 (10): 3098–3103.
Jalava, J., and S. Griffiths. 2017. Philosophers on Psychopaths: A Cautionary Tale in Interdisciplinarity. *Philosophy, Psychiatry, & Psychology* 24 (1): 1–12.
Kelly, D. 2011. *Yuck!: The Nature and Moral Significance of Disgust*. Cambridge, MA: MIT Press.
Killen, M., and J.G. Smetana. 2015. Origins and Development of Morality. *Handbook of Child Psychology and Developmental Science* 3 (7): 701–749.
Langdon, A.J., M.J. Sharpe, G. Schoenbaum, and Y. Niv. 2018. Model-Based Predictions for Dopamine. *Current Opinion in Neurobiology* 49: 1–7.
Lee, S.W., S. Shimojo, and J.P. O'Doherty. 2014. Neural Computations Underlying Arbitration Between Model-Based and Model-Free Learning. *Neuron* 81 (3): 687–699.
Macintosh, N.J. 1983. *Conditioning and Associative Learning*. Oxford: Oxford University Press.
Mahmut, M.K., J. Homewood, and R.J. Stevenson. 2008. The Characteristics of Non-criminals with High Psychopathy Traits: Are they Similar to Criminal Psychopaths? *Journal of Research in Personality* 42 (3): 679–692.
May, J. 2018. *Regard for Reason in the Moral Mind*. Oxford: Oxford University Press.
McGrath, S. 2018. Moral Perception and Its Rivals. In *Evaluative Perception*, ed. A. Bergqvist and R. Cowan, 161–182. Oxford: Oxford University Press.
Montague, R. 2006. *Why Choose This Book? How We Make Decisions*. New York: Dutton.
Moutoussis, M., E.T. Bullmore, I.M. Goodyer, P. Fonagy, P.B. Jones, R.J. Dolan, et al. 2018. Change, Stability, and Instability in the Pavlovian Guidance of Behaviour from Adolescence to Young Adulthood. *PLoS Computational Biology* 14 (12): e1006679.
Nahmias, E., J. Shepard, and S. Reuter. 2014. It's OK if 'My Brain Made Me Do It': People's Intuitions About Free Will and Neuroscientific Prediction. *Cognition* 133 (2): 502–516.
Navarrete, C.D., M.M. McDonald, M.L. Mott, and B. Asher. 2012. Virtual Movrality: Emotion and Action in a Simulated Three-Dimensional "Trolley Problem". *Emotion* 12 (2): 364.
Nichols, S. 2002. How Psychopaths Threaten Moral Rationalism: Is it Irrational to be Amoral? *The Monist* 85 (2): 285–303.
———. 2004. *Sentimental Rules: On the Natural Foundations of Moral Judgment*. New York: Oxford University Press.
Nichols, S., and R. Mallon. 2006. Moral Dilemmas and Moral Rules. *Cognition* 100 (3): 530–542.
Nobes, G., G. Panagiotaki, and C. Pawson. 2009. The Influence of Negligence, Intention, and Outcome on Children's Moral Judgments. *Journal of Experimental Child Psychology* 104 (4): 382–397.

Prinz, J.J., and S. Nichols. 2010. Moral Emotions. In *The Moral Psychology Handbook*, ed. J. Doris, 111–146. Oxford: Oxford University Press.

Rand, D.G., A. Peysakhovich, G.T. Kraft-Todd, G.E. Newman, O. Wurzbacher, M.A. Nowak, and J.D. Greene. 2014. Social Heuristics Shape Intuitive Cooperation. *Nature Communications* 5: 3677.

Rangel, A., C. Camerer, and P.R. Montague. 2008. A Framework for Studying the Neurobiology of Value-Based Decision Making. *Nature Reviews Neuroscience* 9 (7): 545–556.

Redish, D.A. 2013. *The Mind Within the Brain*. Oxford: Oxford University Press.

Rescorla, R.A., and A.R. Wagner. 1972. A Theory of Pavlovian Conditioning: Variations in the Effectiveness of Reinforcement and Non-reinforcement. *Classical Conditioning II: Current Research and Theory* 2: 64–99.

Rossano, F., H. Rakoczy, and M. Tomasello. 2011. Young Children's Understanding of Violations of Property Rights. *Cognition* 121 (2): 219–227.

Schultz, W., P. Dayan, and P.R. Montague. 1997. A Neural Substrate of Prediction and Reward. *Science* 275 (5306): 1593–1599.

Sheffield, F.D. 1965. Relation Between Classical and Instrumental Conditioning. In *Classical Conditioning*, ed. W.F. Prokasy, 302–322. New York: Appleton Century Crofts.

Shenhav, A., and J.D. Greene. 2010. Moral Judgments Recruit Domain-General Valuation Mechanisms to Integrate Representations of Probability and Magnitude. *Neuron* 67 (4): 667–677.

Smetana, J.G., and J.L. Braeges. 1990. The Development of Toddlers' Moral and Conventional Judgments. *Merrill-Palmer Quarterly (1982-)* 36: 329–346.

Somashekhar, S. 2014. Human Rights Campaign Says Barilla Has Turned Around Its Policies on LGBT. *Washington Post*. November 19, 2014.

Sripada, C.S., and S. Stich. 2005. A Framework for the Psychology of Norms. In *The Innate Mind: Volume 2: Culture and Cognition*, ed. P. Carruthers, S. Laurence, and S. Stich, 280–301. Oxford: Oxford University Press.

Strohminger, N., and S. Nichols. 2014. The Essential Moral Self. *Cognition* 131 (1): 159–171.

Sutton, R.S., and A. Barto. 2018. *Reinforcement Learning: An Introduction*. 2n ed. Cambridge, MA: MIT Press.

Tricomi, E., B.W. Balleine, and J.P. O'Doherty. 2009. A Specific Role for Posterior Dorsolateral Striatum in Human Habit Learning. *European Journal of Neuroscience* 29 (11): 2225–2232.

Valentin, V.V., A. Dickinson, and J.P. O'Doherty. 2007. Determining the Neural Substrates of Goal-Directed Learning in the Human Brain. *Journal of Human Neuroscience* 27: 4019–4026.

Van't Wout, M., R.S. Kahn, A.G. Sanfey, and A. Aleman. 2006. Affective State and Decision-Making in the Ultimatum Game. *Experimental Brain Research* 169 (4): 564–568.

Williams, D.R., and H. Williams. 1969. Auto-Maintenance in the Pigeon: Sustained Pecking Despite Contingent Non-reinforcement. *Journal of the Experimental Analysis of Behavior* 12 (4): 511–520.

Yin, H.H., B.J. Knowlton, and B.W. Balleine. 2004. Lesions of Dorsolateral Striatum Preserve Outcome Expectancy But Disrupt Habit Formation in Instrumental Learning. *European Journal of Neuroscience* 19: 181–189.

Chapter 5
Rethinking Moral Motivation: How Neuroscience Supports an Alternative to Motivation Internalism

Chris Zarpentine

Abstract In this chapter, I draw on neuroscientific work to provide support for an alternative account of the relation between moral judgment and motivation. Much recent discussion focuses on the dispute between *motivation internalists*, who hold that there is a necessary connection between moral judgment and motivation, and *motivation externalists*, who deny this. In contrast, I argue that this relation is best seen as a *normative* one: moral judgment *ought* to be accompanied by the appropriate motivation. I support this view by developing a descriptive account of moral psychology informed by research in neuroscience and psychopathology, which I call *affective engine theory*. According to affective engine theory, moral judgment is influenced by two distinct representational systems: *affective mechanisms* and *general representational mechanisms*. This descriptive account supports a *disjunctive conception of moral judgment*, which distinguishes between different kinds of moral judgments on the basis of the influence of different representational systems in their etiologies. I argue that such a conception of moral judgment is both descriptively and normatively adequate. Such considerations provide reasons to reject motivation internalism. However, rather than simply adopt externalism, I argue that the relation between moral judgment and motivation is governed by a normative ideal of moral agency. Such an ideal is required precisely because, as the empirically-informed account of moral psychology defended here makes clear, humans are prone to certain kinds of practical failures that result from a disconnect between distinct representational systems. Together, these arguments demonstrate how research in neuroscience can contribute to normative theorizing: attention to the neuroscientific details highlights the complexity and heterogeneity of moral thought and can therefore guide the construction of normative accounts that are descriptively adequate.

Keywords Moral motivation · Moral judgment · Internalism · Externalism · Affect · Emotion · Moral agency

---

C. Zarpentine (✉)
Associate Professor of Philosophy, Wilkes University, Wilkes-Barre, PA, USA
e-mail: chris.zarpentine@wilkes.edu

G. S. Holtzman, E. Hildt (eds.), *Does Neuroscience Have Normative Implications?*, The International Library of Ethics, Law and Technology 22,
https://doi.org/10.1007/978-3-030-56134-5_5

The goal of this chapter is to draw on neuroscientific work to provide support for an alternative account of the relation between moral judgment and motivation. Many theorists are skeptical that empirical research can be of much use in ethics: ethics focuses on *normative* questions about how things *ought* to be while science can provide insight into *descriptive* matters about how the world *is*. The approach I take here holds that descriptive and normative matters are interconnected. Thus, a descriptive account of the moral psychological mechanisms that underlie moral judgment and motivation, informed by neuroscientific research, can productively inform our normative theorizing.

Much recent philosophical work on the relation between moral judgment and motivation has focuses on arguments for or against various versions of *motivation internalism*. Though formulations differ, the basic idea is that moral judgment has a *necessary* connection to motivation such that an agent who judges that it is morally right (or wrong) to Φ must have at least some motivation to (or not to) Φ.[1] Thus, internalism holds that there is a necessary connection between moral judgment and motivation. In contrast, *externalism* denies that there is a necessary connection.

Here, I argue that empirical work provides support for an alternative to these standard philosophical views. These views account for the relation between moral judgment and motivation in modal terms, either asserting or denying a necessary connection. In contrast to both internalism and externalism, I argue that the relation between moral judgment and motivation is best understood as a *normative* one: moral judgment *ought* to be accompanied by the appropriate motivation. Nevertheless, in practice, there may be many moral judgments where this motivation is absent. As a result, it is possible for an agent to expresses a genuine moral judgment but lack the appropriate motivation and, in such cases, the disconnect will constitute a *moral* failing of the agent.

I proceed by developing a descriptive account of moral psychology. In Sects. 5.1 and 5.2, I draw on research in neuroscience and psychopathology to provide support for this account. This account, which I call *affective engine theory*, holds that two distinct mechanisms are capable of contributing to moral judgment: *affective mechanisms*, specialized for the representation of reward and value, and *general representational mechanisms* that underwrite domain-general discursive representations. However, only affective mechanisms are directly connected to motivational structures. As a result, this descriptive view suggests a *disjunctive conception of moral judgment*, according to which there are different kinds of moral judgments, distinguished by the influence of different representational systems in their etiologies. In Sect. 5.3, I argue that a disjunctive conception of moral judgment is *normatively* adequate. Drawing on these arguments, in Sect. 5.4, I develop a criticism of internalism. Rather than simply endorse externalism, I identify a mistake that underlies the way this debate has been framed. Recognizing this mistake makes apparent the possibility that the relation between moral judgment and motivation is normative. I develop this alternative, arguing that this relation is governed by an ideal of moral

[1] For a recent review, see Björnsson et al. (2015).

agency. I show how the descriptive account defended earlier provides support for this alternative. Ultimately, this demonstrates how research in neuroscience can inform normative theorizing.

5.1 The Neuroscience of Affect and Motivation

While much philosophical work on internalism has proceeded from the armchair, there have been attempts to bring empirical work to bear on this issue. Adina Roskies, for example, has argued that individuals with damage to the ventromedial prefrontal cortex (vmPFC) exhibit normal declarative knowledge and reasoning in the moral domain, yet often lack motivation in "ethically charged situations" (2003, p. 57). Since such individuals appear to make moral judgments without the appropriate motivation, she argues that such individuals represent "walking counterexamples" to internalism (2003, p. 51). One response, by defenders of internalism, has been to question the moral status of individuals with such severe pathologies.[2] If the moral agency of such individuals is sufficiently undermined by their condition, then the putative moral judgments of such individuals are not genuine moral judgments. As a result, such cases cannot represent counterexamples to internalism.

I will not directly address these arguments. Although I will draw on some of the same research that is discussed by Roskies, I will not offer such evidence in support of counterexamples. Instead, I will proceed by developing and defending an account of the mechanisms that underlie moral judgment and motivation by drawing on work in neuroscience. Thus, my arguments will not depend upon contentious claims about the moral status of such individuals and are not subject to this objection.

According to the account I will support here, *affective mechanisms* play a crucial role in both moral judgment and motivation. For this reason, I dub the account *affective engine theory* (AET). The following three claims constitute the primary features of this account:

1. affective mechanisms are structurally distinct from the neurocognitive mechanisms that underlie more general representational capacities—*general representational mechanisms* (GRM),
2. affective mechanisms play a direct role in motivational processes, while GRM only indirectly influence motivational processes, and
3. both affective mechanisms and GRM can directly influence the processes that generate moral judgments.

[2] Kennett and Fine (2008); see also Cholbi (2006).

If correct, AET provides support for a *disjunctive conception of moral judgment* since, according to (3), moral judgments can be produced in at least two different ways and because, according to (2), moral judgments produced in different ways will have different properties: only those moral judgments that are influenced by affective mechanisms will exhibit a necessary connection to motivation. Other philosophers have developed similar views,[3] but none have taken up the specific strategy of appealing to a wide range of empirical research to support a detailed account of the structure of moral psychology. As I discuss below, the neuropsychological details have important consequences for understanding the distinctive properties of the different processes involved in moral thinking. And these descriptive details have implications for normative theories as well.

There is a long tradition in philosophy of grounding morality in sentiments or emotional responses. This tradition is supported by early empirically-informed research in moral psychology (e.g. Haidt 2001; Greene et al. 2001; for a review, see Prinz 2007). In line with recent work in affective science, I use 'affective' to refer to a range of states, including drives (e.g. hunger and thirst), emotions, and moods, which involve *hedonic valence* (i.e. felt pleasure or displeasure) and *arousal* (activation of the autonomic nervous system and endocrine system).[4]

What is emerging from recent research is that such responses depend upon a distinct set of brain regions (including the vmPFC, orbitofrontal cortex (OFC), amygdala, medial prefrontal cortex (mPFC) and regions of the basal ganglia and the midbrain dopamine system).[5] I use the term "affective mechanisms" to refer to this network of neural regions.[6] More importantly, we are gaining a better understanding of the kinds of processing in which these mechanisms are involved. In particular, affective mechanisms implement various forms of associative or contingency-based reinforcement learning, which can be described by substantive, quantitative theories.[7] These learning processes are crucial for the acquisition of physiological responses to stimuli and for updating representations of value that contribute to the production of behavior. As a result, these mechanisms facilitate the acquisition of representations of *reward* and *value*.

Drawing on some of this same research, as well as influential work in psychology on dual process theory (e.g. Stanovich 2009; Kahneman 2011), Peter Railton (2014, 2017) has recently defended a similar account. Describing what he calls "the affective system," he explains it as

[3] See, for example, DePaul (1991), Tolhurst (1995), Campbell (2007), Kriegel (2012), Campbell and Kumar (2012), and Kauppinen (2015).

[4] This draws on "core affect theory," see, for example, Russell (2003), Barrett and Bliss-Moreau (2009), and Barrett and Russell (2014).

[5] For a review of the neuroanatomy of affective processing see Rolls (2005, 2014); see also Kringelbach (2005), Kringelbach and Berridge (2009).

[6] For a critique of this commonly-accepted characterization, see Holtzman (2018).

[7] For further discussion of these learning algorithms and their implementation in the brain, see Rolls (2005, 2014), and Glimcher and Fehr (2014), see also Crockett (2013).

> a system designed to inform thought and action in flexible, experience-based, statistically sophisticated, and representationally complex ways—grounding us in, and attuning us to, reality. It presents to us an evaluative landscape of the physical and social world capable of tacitly guiding perception, cognition, feeling and action in ways anticipated by Aristotle and Kant, among others. (Railton 2014, pp. 846–47)

On this view, affect is a pervasive feature of human cognition.

General representational mechanisms (GRM) can be distinguished from affective mechanisms both by their divergent neurological substrate and differences in the kinds of representations they involve. Research in learning and memory reveals that there is a distinct brain network involving the hippocampi and medial temporal lobes that underlies the capacity for *declarative memory*, which includes *episodic memory* (memory of events) and *semantic memory* (memory of facts) (Shrager and Squire 2009; Spreng et al. 2009). While affective mechanisms involve specialized representations that carry information about reward and value,[8] GRM underwrite general representational capacities. The representations handled by GRM exhibit the stimulus-independence and systematicity that is characteristic of *conceptually structured representations*.[9] For example, I can think about the last time I went on a specific hike without such a thought being prompted by current stimuli (e.g. a perception of the trailhead). Similarly, I can also entertain the possibility of John, Jane, or Jo going on the same hike, consistent with Gareth Evan's "generality constraint."[10] Thus, the conceptual structure of representations handled by GRM gives them broad general representational powers. As a consequence, such representations possess truth-evaluable content and are involved in processes of explicit, logical reasoning.

So far I have offered a fairly standard view according to which human beings possess at least two distinct kinds of representational systems. Drawing on neuroscientific work, I have argued that these systems can be distinguished not only by their reliance on different neural regions, but also by the fact that they involve different kinds of representations. GRM underwrite a domain-general representational system that facilitates the kind of representations that have truth-evaluable content and can enter into processes of logical reasoning. In contrast, affective mechanisms involve representations that are specialized to convey information about reward and value and are involved primarily in the kinds of associative learning processes that update these representations.

The next step is to defend claim (2), that affective mechanisms are directly involved in motivational processes, while GRM influence motivation only indirectly. Going back at least to the eighteenth century, the "affective" has been opposed to both the "cognitive" and the "conative" in the study of the mind.[11] However, according to AET, affective mechanisms play an important role in motivation.

---

[8] For discussion of this kind of account of representation, see Dretske (1988), Prinz (2000, 2002), and Skyrms (2010). For further discussion of the kind of content that affective states possess, see Prinz (2004) and Griffiths and Scarantino (2009).

[9] For further discussion see, for example, Camp (2009).

[10] See Evans 1982, pp. 100–105; for further discussion, see Camp 2009, pp. 276–82.

[11] For an informative historical review of this tri-partite view, see Hilgard (1980).

Indeed, drawing on Timothy Schroeder's *reward theory of desire* (2004), I will argue that some states of affective mechanisms are themselves motivational.

Desire is often considered to be the paradigm of a conative, or motivational, state. While desires are often characterized in terms of their relation to the production of behavior, Schroeder defends an account that defines desires in terms of their role in representing some state of affairs as a *reward* in the context of contingency-based reinforcement learning processes. The kind of learning processes described in Schroder's account are precisely those that are implemented by what I have identified as affective mechanisms. On his view, to desire that *P* is to represent *P* as a reward.

While Schroeder maintains that desires are propositional attitudes, I see no reason to adopt this view.[12] While they do not involve truth-evaluable content or conceptual structure, the representations handled by affective mechanisms are capable of representing the reward value of some object or situation—indeed, these representations are both specialized for this function and are intimately involved in the contingency-based learning processes that are essential to desire on Schroeder's account. Thus, given the role of affective mechanisms in maintaining and updating representations of reward and value, (at least some) desires will depend upon representations of value constituted by states of these affective mechanisms.

In contrast, the representations handled by GRM do possess the conceptual structure necessary for propositional attitudes. However, these representations are not involved in the processes of contingency-based learning but are responsive to the kinds of logical relations that are involved in explicit reasoning. Because these are domain-general representations, they are capable of representing states of affairs as "rewards" in some sense, though not in the sense intended in Schroeder's account. The relevant sense of reward is dependent upon the role of such representations in processes of contingency-based learning. Thus, such general representations cannot be, according to Schroeder's account, desires.

Neuroanatomy provides further support for the view that affective mechanisms are directly involved in motivation. Affective mechanisms are partially constituted by regions of the brain (e.g. the OFC and mPFC) that implement reinforcement learning algorithms. These regions are adjacent to other regions, e.g. lateral prefrontal cortex and anterior cingulate cortex, that are involved in action planning and choice (Glimcher 2009). These latter areas project to motor regions that are involved in the execution of action (Wallis 2007; Kable and Glimcher 2009; Niv and Montague 2009; Wallis and Kennerley 2010).

Thus, neuroscientific work provides support for the claim that affective mechanisms are directly involved in motivation. Schroeder's reward theory of desire offers an account of how certain states of these affective mechanisms actually constitute desires. In contrast, there is no evidence that GRM are directly involved in motivational processes in the way affective mechanisms are. There is, however, evidence that such conceptually structured representations can indirectly contribute to the

[12] For discussion, see Thagard (2006).

production of behavior by influencing affective mechanisms (Hare et al. 2009; Wagner et al. 2013).

Hare et al. (2009), for example, found that the exercise of self-control on the basis of explicit beliefs depends upon the modulation of brain regions that underlie affective mechanisms (e.g. the mPFC) by the dorsolateral prefrontal cortex (DLPFC), an area associated with executive control (Miller and Cohen 2001). Suppose, for example, I am tempted by a piece of chocolate cake. My affective system is representing this as a reward. If, however, as a result of a recent conversation with my physician, I believe that I ought to reduce my caloric intake, then I may try to exercise self-control in this instance. Hare et al.'s study suggests a mechanism by which this occurs: the explicit belief that I ought to watch what I eat exerts influence on my behavior via a process by which the DLPFC modulates areas of the mPFC. In their study, DLPFC activity was inversely related to mPFC activity, suggesting that DLPFC activity was involved in "ramping down" mPFC activity, thereby adjusting the represented value of the cake in light of health considerations.

This provides evidence of an indirect link between explicit beliefs (subserved by GRM) and the states that produce behavior. Many philosophers have maintained that more "cognitive" states are capable of generating motivation of their own.[13] However, when we turn to research in neuroscience, there is no evidence to support the existence of a direct link between GRM and the mechanisms involved in action planning and execution. Rather, the picture that emerges from this research supports the view that the brain regions underlying affective mechanisms (the vmPFC, lateral PFC, and ACC) appear to be a crucial part of a common path to action (Kable and Glimcher 2009; Levy and Glimcher 2012). We only have evidence that GRM can influence behavior indirectly by modulating affective mechanisms.[14] This provides reasonable support for claim (2): in contrast to GRM, only affective mechanisms play a direct role in motivational processes.

So far I have concentrated on providing support for claims (1) and (2): that affective mechanisms and GRM constitute two distinct representational systems and that there is good reason to believe that only affective mechanisms play a direct role in motivational processes. In the next section, I turn to research in psychopathology to provide further support for this account.

[13] For examples, see Mele's discussion of "cognitive engine theory" (2003, Chap. 4).

[14] It is always possible that additional research may undermine such a picture. New work may uncover evidence that supports a more direct link between GRM and motivational mechanisms. Obviously, our knowledge of the brain is not complete. But it seems unlikely that future research will reveal such hitherto unnoticed connections. It is, thus, reasonable to proceed on the basis of what is supported by the evidence currently available.

5.2 Evidence from Psychopathology

In this section, I turn to a discussion of work in psychopathology. I do this in order to provide additional support for AET. However, I also think that research in psychopathology is a crucial source of evidence in support of claim (3): that both affective mechanisms and GRM can directly influence the processes that generate moral judgments. Thus, the arguments of this section are two-fold. First, I will argue that research in psychopathology provides support for this account because endorsing claims (1)–(3) provides the resources for offering a plausible explanation of otherwise puzzling patterns of empirical results.

Second, I will highlight how this evidence is of crucial importance in supporting claim (3), that both affective mechanisms and GRM can directly influence the processes that generate moral judgments, because of a special difficulty that arises in trying to support this claim. This difficulty arises as a result of the representational redundancy of affective mechanisms and GRM posited by claim (1), above. If there are two distinct representational systems, we should expect that both will often be activated in parallel. While this is not surprising, it does pose a problem. If, in neurotypical individuals, distinct representational systems often operate in parallel, it will be difficult to find clear support for the claim that each of these representational systems is independently capable of influencing moral judgment.[15]

Consider, for example, the *Footbridge* case (Thomson 1985), in which participants must judge whether or not it is permissible to push a large man off a footbridge in front of a trolley to his death in order to save the lives of five innocent people stranded on the tracks. Neuroimaging research on such trolley problems finds evidence that consideration of these kinds of cases (when contrasted with more "impersonal" dilemmas) triggers increased activation in areas of the brain associated with emotion (e.g. Greene et al. 2001). However, in ordinary individuals such cases will likely trigger activity in multiple representational systems, making it difficult to isolate the contributions of distinct systems.[16]

It is here that research in neuroscience and psychopathology can be of special assistance. In studying how moral thought and action are affected by the dysfunction of specific mechanisms, such research can provide support for causal claims about how these mechanisms function in normal individuals.[17] As a result, this work can provide crucial support for (3), that both affective mechanisms and GRM can directly influence the processes that generate moral judgment. I begin with a brief

[15] For example, in the literature on learning and memory there is substantial support for the view that both affective mechanisms and a medial-temporal lobe memory system are often engaged in parallel and compete to influence behavior. For discussion, see reviews by Seger (2006), Poldrack and Foerde (2008), and Shohamy et al. (2008).

[16] Greene et al. (2001) recognize this problem and supplement their neuroimaging data with behavioral responses and reaction time results.

[17] For related discussion on the "lesion method," see Koenigs et al. (2007a).

discussion of psychopathy before turning to a more detailed treatment of vmPFC damage.

As I noted in the introduction, there are contentious issues about the moral status of individuals that exhibit moral psychopathology; whether such individuals make "real" moral judgments has been much contested. While I will discuss the moral judgments of psychopaths and vmPFC patients, nothing in my argument is affected if these are thought of as *merely putative* moral judgments. I will use this evidence to support a descriptive account of affective mechanisms and GRM and how each are related to the mechanisms that produce moral judgment. I contend that the very same mechanisms are at work both in cases of psychopathology and in those individuals with no discernable pathology. For ease of exposition, I will often refer simply to the moral judgments of such individuals. The reader is free to imagine this term in "scare quotes" or to retain any appropriate qualifications in the discussion of these pathological cases. When I shift from descriptive to normative issues, I will address similar concerns as they arise.

Psychopathy is characterized by a distinctive combination of *reactive aggression*, a tendency toward violent response to (often minimal) provocation, and *instrumental aggression*, which involves the use of violence as a means toward other ends (Blair et al. 2005, pp. 12–13). At the same time, psychopathic individuals demonstrate a generally normal pattern of moral judgments (Cima et al. 2010; Glenn et al. 2009).[18] Thus, puzzlingly, psychopaths appear to exhibit a normal awareness of moral norms, yet act in ways that violate these norms.

AET provides a plausible explanation for this puzzling pattern of results. Psychopathy results from affective dysfunction, in particular dysfunction of the affective mechanisms that implement reinforcement learning (subserved by the amygdala and vmPFC) (Blair et al. 2005; Blair 2007; Shenhav and Greene 2014). Psychopaths are systematically insensitive to the distress of others and so acquire no aversion to performing actions that harm others. As a result, they fail to acquire many of the affective representations that serve the function of opposing other-harming acts in normal individuals. This deficit disrupts normal moral socialization and accounts for psychopaths' tendency toward instrumental aggression. However, such individuals compensate for this deficit in affective learning by recruiting other, more "cognitive" learning processes subserved by GRM.[19] This compensatory

[18] To be clear, psychopathy may lead to some differences in moral judgment. In particular, there is evidence that psychopaths fail to distinguish moral violations from conventional ones. For a review of some of this research, see Nichols (2004, Chap. 1). However, more recent work has challenged the methodical significance of the moral-conventional task (e.g. Kelly et al. 2007) and raised concerns about whether previously reported effects are an artifact of inappropriate test materials (Aharoni et al. 2012). Moreover, AET can provide an explanation for this apparent difference. See note 20.

[19] This is supported by, among other evidence, differential processing (revealed, for example, by neuroimaging and EEG) on tasks that require processing of affective stimuli, even in cases where psychopaths' behavioral responses are the same. In other words: stimuli that trigger affective processing in normal individuals trigger processing in frontal regions (e.g. the dorsolateral prefrontal

processing explains why, despite a dysfunction of affective learning mechanisms, psychopaths nevertheless exhibit normal patterns of moral judgment.[20]

According to AET, affective mechanisms are distinct from GRM and, while the latter can influence the processes that generate moral judgment, they are not themselves motivational. Thus, on this account, psychopathy results from a dysfunction of affective mechanisms. This deficit leaves GRM intact and these are recruited to compensate for the affective deficit. While these general representations can sustain normal patterns of moral judgment, they are not motivational and the underlying affective dysfunction results in profound effects on social and moral behavior. Since AET offers a plausible explanation of these results, they provide additional support for this account.

Moreover, by demonstrating how GRM can contribute to moral judgment even in the presence of affective dysfunction, this research provides vital support for one half of claim (3): that GRM is capable of influencing moral judgment. In the absence of any reason to believe otherwise, this indicates that GRM can serve this same function in individuals with no discernable pathology. Indeed, there is evidence for similar variability in populations of individuals not suffering from such severe pathology.[21]

Individuals who have suffered damage to the vmPFC present a similar puzzle. They exhibit well-known decision-making difficulties that affect social and moral behavior (Damasio 1994). However, they appear to retain their previously learned social and moral beliefs. This provides further support for claims (1) and (3) since it indicates that affective mechanisms can be damaged while leaving intact the explicit beliefs subserved by GRM and that these latter representations continue to contribute to moral judgment. Moreover, the peculiar behavioral issues that affect such individuals provide additional support for claim (2) by supplying evidence that, despite the persistence of normal patterns of moral judgment, damage to affective mechanisms leads to a significant disruption of the motivational mechanisms that produce social and moral behavior.

So far, this argument closely parallels that offered above in the discussion of psychopathy. However, studies of vmPFC patients on a series of moral dilemmas reveal an instructive and systematic divergence from normal patterns of moral

---

cortex) more associated with higher cognitive capacities in psychopathy. For a review of many of these studies, see Blair et al. (2005, pp. 59–62) and Kiehl (2006, pp. 113–16).

[20] As I discussed above (note 18), there is some evidence that psychopaths do not distinguish between moral and conventional violations in the normal way. Supposing that future research confirms this, the hypothesis that an affective deficit in psychopathy leads to compensatory "cognitive" processing goes some way toward explaining this apparent difference. Since the argument in the text is that AET is supported because it provides a plausible explanation of these results, regardless of the more general significance of these results, they do not undermine the present argument.

[21] On related tasks, a similar pattern of reduced affective processing has been observed in male college students (Gordon et al. 2004), a subset of individuals with substance abuse problems and control groups from these studies (Bechara 2005), as well as individuals who merely demonstrate a tendency toward characteristically utilitarian judgments (Moretto et al. 2010).

judgment: individuals with damage to the vmPFC exhibit a greater tendency toward characteristically utilitarian judgments in dilemmas that pit the direct causing of harm against greater compensating benefits (Koenigs et al. 2007b). The *Footbridge* case is one such dilemma; Joshua Greene and colleagues have dubbed these kinds of cases "personal moral dilemmas" (Greene et al. 2004).

The tendency toward utilitarian judgments reveals something interesting about the function of the vmPFC according to AET: the vmPFC plays a key role in integrating affective information into the decision-making process, via a process that Fellows and Farah describe as an "ongoing, dynamic assessment of relative value" (2007, p. 2673). I dub this process *dynamic evaluation*.[22] Dynamic evaluation can occur in response to *simulation*, the process of imagining future and possible outcomes,[23] as well as *in situ*, in response to continuing experience.

In some of the moral dilemmas used in experimental studies, previously learned moral rules can easily be applied in order to arrive at the (statistically) normal response. Such representations are handled by GRM. These capacities are preserved in vmPFC patients, and such individuals do demonstrate normal moral judgment in many cases. However, consideration of *difficult* personal moral dilemmas[24] requires more than an appeal to straightforward moral rules. In response to such cases, we imagine the situation and rely on our affective responses to this imagining as a guide to decision-making. This is dynamic evaluation in response to simulation. When vmPFC patients are confronted with such dilemmas, damage to the vmPFC disrupts the negative affective response that would normally accompany the thought of pushing a large man in front of a moving trolley. As a result, they are more likely than normal participants to make the characteristically utilitarian judgment and judge that it is permissible to do so.

Perhaps the most unique and puzzling aspect of vmPFC damage involves the inability to act appropriately despite awareness of the relevant features of the situation – even explicit beliefs about what the correct course of action would be. In one study, for example, vmPFC patients were asked to engage in a structured conversation with a stranger. They engaged in socially inappropriate behavior despite an awareness of the social norms they were violating. They were even able to recognize their behavior as inappropriate when shown a videotape of their interaction![25]

As with psychopathy, AET provides a plausible explanation of this otherwise puzzling clinical profile: since affective mechanisms are distinct from GRM, damage to the vmPFC can affect the former and leave the latter intact. GRM allow

---

[22] For further defense of this account of the vmPFC, see Zarpentine (2017).

[23] See, for example, Schacter et al. (2008). Simulation appears to be subserved by the same brain regions (e.g. the medial temporal lobe) that constitute the GRM (Schacter et al. 2009; and Spreng et al. 2009).

[24] Greene et al. (2004, p. 390) identify these cases as ones that generate longer response times, less consensus and generally involve a conflict between overall benefits and otherwise impermissible actions.

[25] Beer et al. (2006). The performance of vmPFC patients on the Iowa Gambling Task provides additional evidence of this (see, for example, Bechara et al. 1996, 1997).

vmPFC patients to recall, entertain, and acquire accurate beliefs relevant to their situation and to make normal moral judgments in a range of cases. However, these representations are disconnected from the motivational mechanisms that produce action. Indeed, Damasio's patient E.V.R. was unable to come to a decision about what to do despite showing normal performance on tasks designed to measure the ability to think of alternative options in a given situation, reflect on the future consequences of actions, engage in means-end reasoning, and foresee the outcome of social situations—tasks that depend upon GRM (Saver and Damasio 1991; Damasio 1994, 49). According to AET, damage to the vmPFC impairs an individual's capacity for dynamic evaluation. In considering hypothetical cases, this can lead to characteristically utilitarian responses to personal moral dilemmas. However, normal individuals also rely on dynamic evaluation *in situ*, and this deficit explains the characteristically inappropriate social behavior and poor decision-making of vmPFC patients.

Thus, the AET gains support because it is able of offer a plausible explanation of otherwise puzzling clinical results. At the same time, however, by revealing the effects of disrupting the function of affective mechanisms, it provides crucial support for claims about the causal influence of GRM. In both vmPFC damage and psychopathy, we see that normal moral judgment is largely preserved. This provides powerful evidence in support of claim (3), that GRM can directly influence moral judgment. However, because of the nature of these pathologies, this evidence provides support only for one half of claim (3): that GRM contributes to the processes that generate moral judgment. There is significant correlational evidence (e.g. from neuroimaging) that affective activation often accompanies moral judgment. Is there any evidence that directly supports the causal claim that affective mechanisms contribute to moral judgment?

A recent study shows that individuals with bilateral hippocampal damage are *less* likely to endorse the characteristically utilitarian option than control participants (McCormick et al. 2016). As noted above, the hippocampi are crucial components of our GRM. In my discussion of vmPFC damage above, I argue that making a characteristically utilitarian judgment in response to difficult personal moral dilemmas depends upon the capacity for dynamic evaluation in response to simulation. According to AET, damage to GRM would disrupt the capacity for simulation and, in the absence of such processing, we would expect judgment to be driven by negative affective responses to actions that directly cause harm to an individual, making individuals less likely to endorse the characteristically utilitarian judgment in such cases. Again, AET provides a plausible explanation for the pattern of results we observe in this case of psychopathology. When the capacities subserved by GRM are disrupted by damage to these mechanisms, we should expect individuals' responses to be dominated by their affective responses. This is exactly what we observe. Thus, this research provides direct support for the independent contribution of affective mechanisms in moral judgment.

Experimental manipulation studies are also important in providing evidence of the causal contribution that affective mechanisms make to moral judgment. Greene et al. (2008) used a cognitive load task, which selectively interferes with "cognitive" processing without disrupting affective processing, to shed light on the contributions of different kinds of processes. The results indicate that while cognitive load increased the response time for characteristically utilitarian judgments, it did not increase response time for other judgments. This provides some evidence for the independent influence of affective mechanisms on moral judgment in some cases.

Experimental affective induction can also provide evidence for the causal influence of affective mechanisms on moral judgment. Surreptitiously inducing an affective response can produce a measurable effect on moral judgment. For example, disgust induced by hypnosis has been reported to increase the severity of moral judgment and, in some cases, to lead to a negative moral judgment of an otherwise innocuous action (Wheatley and Haidt 2005). Other instances include manipulation by novelty fart spray, unpleasant environments or disgusting video clip (e.g. Schnall et al. 2008[26]; Valdesolo and DeSteno 2006). Interestingly, these sorts of manipulations failed to influence moral judgment when they were *too obvious*: when disgust was rightly attributed to the experience of submerging one's hand in a bucket full of a gooey substance, there was no effect on moral judgment (Schnall et al. 2008). This suggests that manipulations are successfully only when they are able to trigger an affective response without triggering the acquisition of a corresponding explicit belief subserved by GRM.[27] While less robust than the evidence in support of the independent influence of GRM on moral judgment, I believe there is reasonable support for the claim that affective mechanisms can independently influence moral judgment.

Research in psychopathology reveals puzzling patterns of clinical results. AET gains support from the fact that it can provide a plausible explanation of these puzzling results. Moreover, in combination with experimental manipulation studies, this work provides crucial support for claims about the independent causal influence of affective mechanisms and GRM on the processes that produce moral judgment.

5.3 A Disjunctive Conception of Moral Judgment

While the above discussion has not been exhaustive, I believe it provides strong support for the three central claims of AET: that affective mechanisms and GRM are distinct and while only affective mechanisms play a direct role in motivational processes, both can influence the processes that produce moral judgment. This account is a descriptive one. However, given the approach adopted here, we should expect a close connection between this account and any plausible normative account. Taking this descriptive account as a starting point suggests a *disjunctive conception of moral judgment* (DMJ) according to which there are at least two kinds of moral

[26] Editors' note: Schnall et al.'s (2008) research has failed to replicate (see, e.g., Ugazio et al. 2012).

[27] This may help explain why, in contrast to finding evidence of the influence of GRM on moral judgment, it is relatively more difficult to find evidence supporting the independent contribution of affective mechanisms on moral judgment.

judgment: moral judgments that are influenced by affective mechanisms, which will entail the appropriate motivation, and moral judgments that depend only on GRM, which will not. In this section, I shift the attention to normative issues and argue that DMJ provides a normatively adequate account of moral judgment.

One source of support for DMJ comes from reflection on ordinary moral experience. When we do so, we often find a multitude of things going on during moral decision-making, including, *inter alia*: the consideration and application of explicit principles, reflection on our affective responses to imagined possibilities (i.e. dynamic evaluation in response to simulation), even reliance on "gut-reactions," which we may have difficulty explaining or justifying, but nevertheless reject only with great difficulty. Indeed, I think it is quite plausible that, more generally, moral agency requires a host of cognitive capacities that rely on both affective mechanisms and GRM. In the absence of our affective responses to moral situations, we would lack the phenomenology that allows us to conceive of moral demands as normative; without GRM we would be unable to step back from these affective responses and to engage in the explicit reasoning necessary to endorse them or reject them. The DMJ captures the inherent complexity and heterogeneity that we find in moral psychology. At the same time, it accommodates the possibility that, on any given occasion, moral judgment may be independently influenced by either affective mechanisms or GRM.[28]

Still, when we turn from an account of how people *do* make moral judgments to an account of how they *ought* to, there may be reasons to adopt a more restrictive account of moral judgment. Thus, it pays to focus some critical attention on more restrictive accounts of moral judgment. Some philosophers, especially those in the sentimentalist tradition, may consider emotions or affective representations to be the only legitimate source of moral judgment. Call such a view *affective essentialism* about moral judgment. Others, especially philosophers more inclined toward rationalism, may wish to privilege the role of explicit reasoning (subserved by GRM) in generating moral judgments. Call this *explicit cognitive essentialism* about moral judgment—*cognitive essentialism* for short. I consider each of these in turn.

One of the most well-developed examples of affective essentialism appears in the work of Jesse Prinz (2007, 2015). Prinz draws from research in psychology and neuroscience on the central role of emotional processes in moral judgment in developing a view he calls *emotionism*. On the basis of this research, he defends the view that emotion is necessary and sufficient for moral judgment. Given the motivational properties of emotions, Prinz endorses a version of motivation internalism. While he

[28] Some "hybrid" accounts of moral judgment hold that it is a complex state involving both cognitive (e.g. belief) *and* conative (e.g. desire) or affective (e.g. moral emotion) aspects (e.g. Copp 2001; Ridge 2006; Campbell 2007; Campbell and Kumar 2012). However, as Campbell (2007, pp. 338–40) argues, neither Copp's *realist-expressivism* nor Ridge's *ecumenical expressivism* is genuinely hybrid since they privilege the cognitive or expressive aspect, respectively. Campbell writes: "In a fully hybrid theory both parts of moral judgment, the belief and the state of emotion and motivation, have equal status, to the extent that each can function by itself as a moral judgment in the absence of the other" (2007, p. 340). This is consistent with DMJ as articulated in the text.

recognizes that other processes are involved in moral thought, he doesn't assign them an essential role.

The problem for such accounts is that they neglect the important role that explicit reasoning can play in moral judgment. Consider the kind of moral thinking that philosophers often engage in: after committing to a specific moral principle, one proceeds to consider how it applies to a particular case at hand and issues a moral judgment accordingly. Affective essentialism would seem to rule out putative moral judgments based on such reasoning since they will not necessarily involve affective mechanisms. Indeed, on such a view, considerations based on explicit reasoning would lack the justificatory standing to contest the moral judgments that derive from affective mechanisms. Since explicit reasoning is not a legitimate source of moral judgment, the output of such a process could not conflict with affective representations on moral matters. In short: if it is true that an important aspect of moral agency is the ability to step back from our affective responses and to consider reflectively whether to endorse them or not, then (at least in some cases) this reflective endorsement or refusal would itself constitute a moral judgment. But this would be impossible according to affective essentialism. In failing to capture this important aspect of moral thought, affective essentialism fails to provide a normatively adequate account of moral judgment.

On the other hand, consider explicit cognitive essentialism about moral judgment. Such a view may be rarely overtly endorsed. Nevertheless, I think many externalists and some rationalists tacitly subscribe to such a position.[29] Regardless, when one considers the descriptive account informed by neuroscience and psychopathology, cognitive essentialism offers a natural way of defending a more restrictive normative account of moral judgment. Thus, it is instructive to consider the problems such a view faces.

One issue concerns the normative demandingness of moral judgments. Above, I suggested that this normativity derives (at least in part) from the phenomenology of affective responses. C.L. Stevenson, for example, speaks of the "magnetism" of ethical terms, which he understands as "the connection between goodness and actions" (1937, 27). The experience of affective responses (and, in particular, their motivational properties) goes a long way toward explaining this phenomenon. Cognitive essentialism, in contrast, has difficulty recognizing this important aspect of moral thought. Moral thought, on such a view, would resemble mere *categorization*, becoming akin to sorting blocks by color or shape. This does not do justice to moral experience. Indeed, cognitive essentialism cannot account for the moral importance of the kind of *evaluative perception,* discussed above in the context of

[29] Peter Singer (2005), for example, comes close to endorsing this view when, in a discussion of the work of Greene and colleagues, he writes: "...Greene's research suggests that in some people, reasoning can overcome an initial intuitive response. That...seems the most plausible way to account for the longer reaction times [of some subjects in response to the footbridge example, and] the preliminary data showing greater activity in parts of their brain associated with cognitive processes...Moreover, the answer these subjects gave is, surely, the rational answer" (2005, 350).

Railton's description of the "affective system" (2014) and which, according to AET, arises from the activity of affective mechanisms.

When we are doing philosophy, we are particularly prone to assuming that moral thinking is all and only about the consideration of explicit arguments or the articulation of systems of values. In the grip of such an assumption, cognitive essentialism may seem plausible. However, the underlying assumption should be rejected. Perhaps more than any other philosopher, Cora Diamond has forcefully argued against the view that the only way to rationally convince someone is through explicit argument.[30] She points to both literary and philosophical work whose aim is to exercise or expand our moral imagination and to encourage us to attend to the world in a certain way. When such works are successful, the process involves coming to have certain kinds of emotional responses to the world.

As examples, she considers some of the novels of Charles Dickens in which one of his aims is to encourage his readers to come to have a greater concern for children. As Diamond puts it: "to enlighten the understanding and ameliorate the affections by providing descriptions which stimulate imagination and moral sensibility" (1991, 299). This is not just presenting the reader with descriptive facts; rather, such writing "expresses a particular style of affectionate interest in human things and imaginative engagement with them" (1991, 300). Exclusive focus on explicit reasoning would delegitimize such strategies for convincing others to share our moral concerns: it would undermine the importance of encouraging the cultivation of the affective responses that are necessary to attend to the world in the right sort of way.

Diamond develops these considerations from the armchair. However, it is an additional strength of the empirically-informed account articulated above that it coheres well with such arguments. In developing this account, I highlighted the importance of dynamic response to simulation – a process of reflective imagining that involves attending to our affective responses. This is precisely the sort of "imaginative engagement" that is prompted by the right kind of literary works and which Diamond is encouraging us to recognize, and more fully appreciate the importance of, in our moral reasoning.

Similarly, reflection on our own moral experience can provide an important counterpoint to the assumption that moral thinking involves only the articulation and evaluation of explicit arguments. When we attend to the experience of making a difficult moral decision we find a host of capacities at work. Consider, for example, a decision about how to treat a student guilty of plagiarism. We might consider general moral principles about honesty and fairness, while at the same time recognizing that impartially applying the policy on our syllabus is not straightforward. To what extent was the offense intentional, as opposed to careless? Was the student desperate, apathetic or just arrogant? One may consider how the student is likely to respond, especially if one is concerned to resolve the situation in a way that recognizes the seriousness of the misconduct but also, if it seems appropriate, allows the

[30] See, for example, Diamond (1982), reprinted in her (1991), page numbers refer to the later. See, also, DePaul (1991), for independent development of a similar argument in the context of debates about internalism.

student an opportunity to learn from the experience. It is hard to imagine proceeding in such a case without allowing ourselves to be informed by, and to take seriously as part of this deliberative process, a range of affective responses that we experience throughout this process. Together, these considerations provide a strong case against cognitive essentialism.

I began this section by discussing how, when we shift from descriptive issues to normative ones, the empirically-informed account of moral psychology defended in the first two sections suggests taking the disjunctive conception of moral judgment as a starting point. The DMJ not only fits well with AET, it also coheres well with reflection on ordinary moral experience. Furthermore, I have raised some serious problems for more restrictive normative accounts of moral judgment. In particular, I have tried to highlight how this empirically-informed account compliments the sorts of arguments accessible from the armchair.

Indeed, reflection on ordinary moral experience and attention to the neuroscientific details both provide independent support for the view that moral thought involves the complex interaction of distinct processes. Such considerations require a normative account of moral judgment that captures this complexity. To be clear, I do not aim to rest this argument on a kind of naïve *Panglossianism*, according to which *however* moral thinking happens to work is normatively ideal.[31] Rather, my claim is that, when we understand the complexity of human moral psychology, we recognize the value of structuring such a system in this way. As a result, normative accounts that impose significant restrictions on the proper functioning of this complex system face serious problems. Thus, when we shift from descriptive concerns to normative ones, we find good reasons to endorse a disjunctive conception of moral judgment.

5.4 A Theoretical Alternative

Suppose, in light of the arguments of the previous section, we take the disjunctive conception of moral judgment to be normatively adequate, where does this leave us when it comes to the relation between moral judgment and motivation? As I noted in the introduction, much recent philosophical discussion about the relation between moral judgment and motivation focuses on arguments for or against some form of *motivation internalism*, according to which moral judgment necessitates (at least some) motivation to act accordingly. In this section, I will argue that the empirically-informed account of moral psychology defended above (AET) together with a disjunctive conception of moral judgment (DMJ) provide the resources not only to criticize extant views, but to provide support for an alternative account of the relation between moral judgment and motivation.

---

[31] For further discussion of Panglossianism in the context of evolutionary theory, see Dennett (1995).

According to AET, there is significant potential for functional redundancy in human moral psychology. As I note above, both affective mechanisms and GRM will very often be activated in parallel. As a result, in individuals without serious pathology, there may very rarely be clear counter-examples to internalism. That is, it may only rarely happen that an individual makes a moral judgment without at least some motivation to act accordingly. This claim alone, however, does not vindicate internalism. What is at issue in this discussion is not whether, as a matter of fact, individuals who make moral judgments are motivated to act accordingly but whether it is *necessarily* the case that an individual who makes a moral judgment is motivated to act accordingly.

It is here that AET and DMJ raise a problem for internalism. According to AET, GRM can independently influence the mechanisms that produce moral judgment. Thus, it is possible for an individual to make a moral judgment without any activation of affective mechanisms. If AET is correct and GRM possess only an indirect link to motivational structures, then it will follow that it is possible for an individual to make a moral judgment, grounded in such general representations, *without* necessarily having any motivation to act accordingly. Since this is a possibility, a connection between moral judgment and motivation cannot be *necessary*. Thus, internalism is false.

It is important to note two things about the foregoing argument. First, as I noted earlier, this argument does not depend upon contentious claims about the moral status of individuals with significant psychopathology. The discussion of psychopathy and vmPFC damage was introduced to defend AET, a descriptive account of moral psychology. I suggested that this descriptive account created a presumption in favor of DMJ, but my arguments that DMJ is a normatively adequate account of moral judgment did not depend upon a discussion of these pathological cases. To reiterate: I relied on research in psychopathy to defend a descriptive account of moral psychology. However, the normative adequacy of DMJ does not presuppose that, for example, vmPFC patients make "real" moral judgments. Indeed, DMJ appears to be neutral on this issue.[32]

Second, the disjunctive conception of moral judgment holds that there are different kinds of moral judgments. The argument just bruited against internalism holds that only moral judgments that are produced through the influence of affective mechanisms will exhibit a necessary connection to motivation. To see more clearly what this argument does (and does not) demonstrate, it will be useful to distinguish between two forms that internalism might take.[33] On the one hand, internalism

---

[32] Elsewhere, I argue that it is plausible to consider the moral agency of vmPFC patients to be partially impaired (Zarpentine 2017, pp. 243–46). Given this sort of impairment, it remains open whether some, all or none of the moral judgments of vmPFC patients ought to be considered genuine.

[33] For further discussion of this sort of distinction, see Tresan (2006, 2009; cf. Little 1997). Tresan distinguishes these two kinds of internalism in terms of the specific form of necessity they involve: *de re* versus *de dicto*. For this reason, in the text, I refer to these views as *metaphysical internalism* and *semantic internalism*, respectively.

might require that genuine moral judgment express a single state that has both representational content and motivational force. Because it combines aspects of belief and desire, such a state is sometimes called a "besire" (Altham 1986; see also Smith 1994). Indeed, this understanding of internalism best captures the *internal* connection between moral judgment and motivation that prompted the view to be so-called, i.e. the sense in which, as William Frankena put it, "motivation is somehow to be 'built into' judgments of moral obligation" (1958, p. 41). Call this form *metaphysical internalism.*

On the other hand, internalism may be taken not to require some "internal" connection between moral judgment and motivation, but merely to claim that any putative moral judgment made without the accompanying motivation simply fails to qualify as a genuine moral judgment. Moral judgment need not *include* motivation, but it must, necessarily, be accompanied by it. Call this variety *semantic internalism.*

With this distinction in mind, it should be fairly clear that the argument articulated above is most effective against metaphysical internalism. It shows that moral judgment may derive from a state that is not, itself, motivational. While DMJ, together with AET, do militate against semantic internalism, additional considerations need to be marshaled to support a robust critique.[34] Still, I find it quite implausible to simply insist that any judgment lacking a necessary connection to motivation is, *ipso facto*, not a moral judgment; the argument offered so far casts enough doubt on internalism to warrant considering theoretical alternatives. Moreover, once my preferred alternative has been introduced, I'll be in a position to offer one additional argument that more specifically targets semantic internalism.

If the considerations discussed so far raise problems for internalism, it might be thought that it is necessary to endorse *externalism.* At a certain level, this is unavoidable. If one argues against a necessary connection between moral judgment and motivation, then one must, presumably, accept that the relation is purely contingent. This is fine as far as it goes. However, in accepting such a characterization, externalism risks collapsing into the view that some people just happen to be motivated to do the right thing. This makes it far too easy to caricature externalism, as Michael Smith does. He suggests that, when one adopts such a view, one becomes committed to the idea that "…whether or not people who have a certain moral belief desire to act accordingly must now be seen as a further and entirely separate question. They may happen to have the corresponding desire, they may not. However, either way, they cannot be rationally criticized" (Smith 1994, 9). This way of putting things seems to place moral thought on a par with choosing a flavor of frozen yogurt or a trim package on a new car. This, too, seems an inadequate way of understanding moral thought.

If the question is simply whether moral judgments necessarily exhibit an internal or necessary connection to motivation, then the arguments put forward to this point support a negative answer. However, I believe that attention to the neuroscientific details of human moral psychology indicates that there is more to be said. In fact, I

[34] Elsewhere, I offer a more sustained criticism of semantic internalism (Zarpentine n.d.).

think the mistake lies in the way this debate has been framed: the opposition of internalism and externalism focuses attention on the relation between moral judgment and motivation purely in terms of the modalities of necessity and possibility. Internalists assert a necessary relation; externalists deny this. These modalities do a poor job of handling the complexity and heterogeneity of moral thinking and motivation that is revealed by careful attention to neuroscientific research. The idea to be developed and defended in the remainder of this chapter is that the connection between moral judgment and motivation is best understood as a *normative* and, in particular, a moral one. This discussion reveals how research in neuroscience can inform normative theorizing.

P.T. Geach, writing about 'good,' identifies the problem that underlies the mistaken way in which this debate has been framed. He labels it a "crude empiricist fallacy." He writes:

> Even if not all A's are B's, the statement that A's are normally B's may belong to the *ratio* of an A. Most chess moves are valid, most intentions are carried out, most statements are veracious; none of these statements is just a rough generalization. (1956, p. 39)

The point is, in each of these cases, limiting the discussion to whether the relation between two things is either necessary or purely contingent obscures the fact that the relation between them is governed by a constitutive norm. If it were not the case that *most* chess moves are valid, there would be no such thing as chess; if it were not the case that most intentions are carried out, the notion of an "intention" would break down. My claim is that framing the debate about the relation between moral judgment and motivation as a dispute between internalism and externalism makes the same mistake. Recognizing this mistake allows us to see an alternative way of approaching the issue.

Specifically, I claim that the relation between moral judgment and motivation is governed by a substantive, constitutive norm of moral agency: *other things being equal, moral judgment* morally ought *to be accompanied by motivation to act accordingly*.[35] The "ought" at issue is a specifically *moral* ought. But, in contrast to *practical rules*, which aim at specifying the *actions* one ought to perform, this norm focuses on the kind of *agent* one ought to be. As a result, this *agential ideal* holds that among the diverse kinds of processes and mechanisms that are involved in moral thought and agency, there are certain sub-personal relations that ought to obtain. Thus, rather than being a matter of metaphysical or semantic necessity, the relation between moral judgment and motivation is *normative*: it is governed by a substantive ideal of moral agency. Such an ideal is as important to morality as

[35] It should be noted that the *ceteris paribus* clause is not trivial, there will be some cases in which other things are not equal and thus there is no requirement that a moral judgment be accompanied by motivation. In some cases, what one ought to do is to reconsider one's moral judgment rather than to bring it about that one is motivated to act accordingly. Moreover, the type of normativity at issue is not the normativity of rationality but, as I note in the text, is a specifically *moral* normativity. As a consequence, the view I defend does not simply collapse into rationalist versions of internalism as articulated, for example, by Michael Smith in his "practicality requirement"(Smith 1994, pp. 61–62). Thank you to an anonymous referee for prompting me to clarify this point.

following the rules of chess is to chess. It is the difference between playing chess and simply "taking" the pieces of one's opponent willy-nilly.

To illustrate how this account differs from internalism and externalism, consider how each of these views would respond to the following case:

> *Bill*: Taking an introductory philosophy class, Bill considers a fairly standard utilitarian argument for vegetarianism. He confirms that the argument is valid. He is already a committed utilitarian and finds no flaws in the other premises. As a result, he endorses the conclusion, and makes the moral judgment that (under most ordinary circumstances) it is morally wrong to eat meat. After class, another student, Sue, asks whether he plans to give up eating meat. Bill demurs. Sue responds, "You mean you don't think the argument is sound?" Bill replies, "No, I think it is sound alright. I continue to endorse the conclusion that it is morally wrong to eat meat. But I don't have any motivation to act accordingly." Bill has made a moral judgment but lacks the motivation to act accordingly.

On its face, such a case seems possible. Externalists will likely see no reason to doubt Bill's sincerity either with respect to his moral judgment or his lack of motivation. Indeed, externalists may find it somewhat commonplace—it is quite similar to the kinds of cases that are often used to motivate externalism. Still, something seems to have gone wrong.

Because Bill is described as both making a moral judgment and lacking any motivation to act accordingly, internalists will want to claim that the case has been mistakenly described. According to metaphysical internalism, the case is impossible: given that moral judgment necessarily depends on a state that involves both doxastic and motivational features, Bill cannot have made the moral judgment and completely lack motivation. In such a case, Bill must be understood as making some kind of mistake: either he failed to make a real moral judgment or he has some, perhaps unnoticed, measure of motivation.

The semantic internalist, too, will take issue with the case. While it is more understandable on this view how Bill could make the apparent moral judgment that it is wrong to eat meat while completely lacking motivation to act accordingly, he is nevertheless mistaken. Because he completely lacks motivation, according to semantic internalism, his apparent moral judgment cannot be a genuine moral judgment. Thus, his error may lie in a misunderstanding of the concept of moral judgment itself.

In contrast, the agential ideal I introduced above identifies the problem as a specifically moral one. Bill has violated an ideal of agency. To be clear, it is important to note that the case does not describe whether Bill goes on to actually eat meat. If he does, then his action may or may not be morally wrong—this will depend upon the moral status of meat eating. If Bill's new judgment that eating meat is morally wrong is correct and he goes on to eat meat, he would be knowingly violating a practical rule. However, *regardless* of whether he goes on to do so (indeed, regardless of whether he is even correct to judge that it is morally wrong to eat meat), by failing to have the appropriate motivation he is violating the agential ideal. What the agential ideal requires, when he finds himself in such a situation, is that he work toward resolving the apparent conflict: this may involve critically re-examine the argument that led him to make this moral judgment in the first place or it may involve trying to bring it about that he is motivated to act accordingly.[36] On this view, what has gone wrong in this case is that Bill seems impervious to the agential ideal he is violating and this is a particularly *moral* failure.

It is not my intention here to beg the question against the internalist (or the externalist). I offer the discussion of this case purely for the purposes of illustration. However, I do, now, want to provide some arguments in support of the agential ideal. Here, I focus on two considerations in favor of this account. I will first discuss how the neuroscientifically-informed descriptive account of moral psychology defended above lends support to this view. I will then argue that the agential ideal provides a better treatment of some putative counter-examples put forward by externalists.

Above, I drew on research in neuroscience and psychopathology to support a descriptive account of moral psychology that identified two distinct representational systems, with different representational powers and relations to motivation. I went on to argue that this complexity is a crucial feature of our moral psychology and that this ought to lead us to accept a disjunctive conception of moral judgment as normatively adequate.

While there are good reasons for the kind of complexity our moral psychology exhibits, a consequence of this complexity is the potential for conflict between distinct representational systems. This leaves us vulnerable to the kinds of practical failures that result from such conflicts. We may, as a result of our explicit beliefs, make certain moral judgments. But there are a number of ways in which we may fail to have the motivation to act accordingly. As Michael Stocker puts it in his classic discussion:

> Through spiritual or physical tiredness, through accidie, through weakness of body, through illness, through general apathy, through despair, through inability to concentrate, through a feeling of uselessness or futility, and so on, one may feel less and less motivated to seek what is good. (1979, 744)

In such cases, our explicit beliefs about what is good or worth pursuing do not entirely match our current stock of motivations. Stocker highlights only some of the circumstances that can lead to such a situation. Sometimes these depend upon psychopathology, sometimes not. Some such conditions are short-lived while others are chronic. Some are relatively innocuous, others are more serious. Indeed, the peculiar pattern of symptoms that exist in cases of psychopathology arise from the most extreme kind of dissociation between the distinct representational systems that constitute our moral psychology. But, the structure of our moral psychology makes all of us, even individuals without serious pathology, prone to such dissociations.

Self-deception is one kind of practical failure that results from such dissociation; self-deception involves the persistence of conflicting representational states. Recently, Daniel Batson and his colleagues have found evidence for widespread

[36] Of course, the agential ideal also recommends that agents be proactive to avoid finding themselves in situations like Bill's. However, few of us are so perfect (or so lucky) as to never find ourselves in such a predicament.

self-deception in ordinary moral thought.[37] When we recognize how widespread and commonplace such practical failures are and that the structure of our moral psychology is what makes us particularly prone to such failures, it becomes clear why the agential ideal is necessary. No factors or forces external to the practice of morality will guarantee that conflicts between different representational systems are resolved. Indeed, Batson's research raises the distinct possibility that moral self-deception is actually *adaptive* from an evolutionary perspective. From *certain* perspectives, other kinds of practical failure may also be desirable. I contend that the pressure on agents to work toward greater sub-personal coherence stems from a specifically moral requirement: an agential ideal according to which, other things being equal, moral judgment ought to be accompanied by motivation to act accordingly.

If our moral psychology did not make us prone to such practical failures, we would have no need for such an agential ideal. Thus, the argument here draws on the sort of point made by Philippa Foot in conceiving of the virtues as *correctives*. On her view, courage and temperance are virtues because humans are particularly susceptible to temptation by fear or desire (2002, 8–10). Similarly, we need the agential ideal because we are particularly prone to the practical failures that arise from the dissociation between different representational systems in our moral psychology.

There is a second consideration that provides support for the agential ideal. This account offers a more natural and plausible treatment of the kinds of putative counter-examples that are contested between internalists and externalists. Consider, for example, Russ Shafer-Landau's example of cowardice in which concentration on the risks of doing one's duty in wartime sap all motivation to act accordingly (2003, 149–50). He intends this as a counter-example to internalism. Above, in my discussion of the *Bill* case, I noted the sorts of responses to such putative counter-examples that internalists tend to offer. Again, that discussion was purely illustrative. Now, however, I want to criticize the kinds of responses to such cases that internalists seem committed to providing.

In response to Shafer-Landau's case, a metaphysical internalist must hold that it is impossible as described. As a consequence, either the individual must be mistaken about judging what his duty is or he must be mistaken about his complete lack of motivation. It is impossible for an agent to have made a genuine moral judgment about what duty requires and completely lack motivation to follow through on that judgment. For a semantic internalist, the response must be similar. The only difference is that, rather than identifying it as a metaphysical impossibility, the semantic internalist can grant that the agent has a belief about his duty. However, semantic internalism holds that the concept of moral judgment precludes the possibility that an individual can make a genuine moral judgment without the appropriate motivation. Thus, the mistake can be attributed to a kind of conceptual misunderstanding of moral judgment.

[37] For a summary, see Batson (2008).

I am not denying that individuals can make mistakes of this kind. However, notice how *peculiar* these sorts of mistakes are. Imagine this soldier's commanding officer (CO) adopting such a view – rather than offering a rousing patriotic speech meant to engage the errant soldier's affective sensibilities in an attempt to motivate this soldier to disregard the risks and march into battle, the CO offers him a lecture of the concept of moral judgment, hoping to correct the unfortunate misunderstanding that stands in the way of his appropriate motivation.

Obviously the CO is more concerned about getting the soldier on the battlefield than with a proper understanding of the relation between moral judgment and motivation. Still, it is helpful to contrast the kinds of responses that an internalist must give with what seems like the most natural treatment of this case. In this case, the most straightforward diagnosis is simply that the soldier fails to have the *courage of his convictions*. He seems clear about what his duty is, but he lacks the motivation to do his duty. His problem is not some esoteric metaphysical or conceptual misunderstanding—it is a *moral* failure. Specifically, he is violating the agential ideal: based on his beliefs about his duty, he is exhibiting cowardice—a failure of character. He is failing to be the kind of agent he ought to be. It may not be in his power to immediately become the kind of agent he ought to be based on his sincere beliefs. Nevertheless, he ought to proceed in a way that aims to resolve this conflict, perhaps by trying to bring it about that he comes to have the appropriate motivation or, perhaps, re-examining what he takes his duty to be—his reluctance in the face of danger may, in fact, be a sign of his uncertainty about the justice of the cause. The agential ideal holds that the conflict ought to be resolved, though it does not necessitate that it be resolved in one way rather than another.

Internalists are right that *something* has gone wrong in such cases. However, neither metaphysical internalism nor semantic internalism seems to offer a plausible diagnosis. This gives us yet another reason to reject both metaphysical internalism and semantic internalism. The agential ideal, on the other hand, gains additional support from the fact that it coheres well with what seems like the most natural treatment of such a case. In Shafer-Landau's case, the problem seems to be a particularly moral failing—the agential ideal provides a specification of this moral failing. I contend that this account has the ability to offer plausible interpretations of many other such disputed cases.

Before concluding, it will be helpful to address one concern about the theoretical alternative I have defended. One might worry that this view depends upon a commitment to virtue ethics. And, since many philosophers favor ethical theories whose focus is on moral principles, this commitment may be seen as problematic. Space does not permit a full discussion of this concern; however, a few comments are in order.[38]

I have argued that the best way of understanding the relation between moral judgment and motivation is in terms of an agential ideal according to which, all

[38] For fuller treatment of a similar problem, see Bok (1996, p. 190 ff.), upon whose ideas the discussion here draws.

other things being equal, moral judgment morally ought to be accompanied by motivation to act accordingly. This does imply that any adequate ethical theory must include the disposition to meet this ideal as a virtue. However, on its own, this agential ideal leaves open what the source of true moral judgments is. This ideal is consistent with a variety of ethical theories, ranging from those that assert a single fundamental moral principle to the most extreme forms of particularism. The view does maintain that the correct ethical theory cannot be limited to the articulation of principles. Arguably, however, the thought that the scope of ethical theory is exhausted by the articulation of principles is plausible only when one is under the spell of the assumption that moral thinking is all and only about the consideration of explicit arguments. In Sect. 5.3, I have argued that we should reject this assumption. Having done so, it is unclear whether there are any good reasons to maintain such a restricted conception of ethical theory. Thus, accepting the normative account of the relation between moral judgment and motivation, for which I have argued here, commits one only to the claim that any plausible ethical theory must recommend at least one virtue. But this is a claim that we have good independent reasons to accept. And doing so does not preclude adopting an ethical theory that is primarily focused on principles.

5.5 Conclusion

In this chapter, I have aimed to reframe the debate about the relation between moral judgment and motivation and to provide support for a theoretical alternative. I have drawn on research in neuroscience and psychopathology to support a descriptive account of moral psychology: *affective engine theory*. Partly on the basis of this descriptive account, I have argued for the normative adequacy of a disjunctive conception of moral judgment. Drawing on these arguments, I offered a critique of internalism. Rather than endorse externalism, I identified a problem that underlies the way this debate has been framed. Recognition of this problem suggested an alternative. I developed and defended this alternative, arguing that the relation between moral judgment and motivation is best seen as a normative one. Specifically, I argued that this relation is governed by an agential ideal, according to which, other things being equal, moral judgment *morally ought* to be accompanied by the appropriate motivation. On this view, an individual may make a moral judgment without the accompanying motivation. However, such cases will constitute a moral failing of the agent. Drawing on the descriptive account informed by neuroscientific work defended earlier, I supported this alternative by appealing to the way the agential ideal entreats us to overcome the kinds of practical failures to which we are particularly prone as a result of our moral psychology. I also argued that this account offers a more a plausible treatment of the kinds of cases that have been the subject of prolonged disputes between internalists and externalists. In the process, I have tried to show one way in which research in neuroscience can make significant contributions to normative theorizing.

References

Aharoni, E., W. Sinnott-Armstrong, and K.A. Kiehl. 2012. Can Psychopathic Offenders Discern Moral Wrongs? A New Look at the Moral/Conventional Distinction. *Journal of Abnormal Psychology* 121 (2): 484–497.

Altham, J. 1986. The Legacy of Emotivism. In *Fact, Science, and Morality: Essays on A.J. Ayer's Language, Truth, and Logic*, ed. G. MacDonald and C. Wright. Oxford: Blackwell.

Barrett, L.F., and E. Bliss-Moreau. 2009. Affect as Psychological Primitive. In *Advances in Experimental Social Psychology*, ed. M.P. Zanna, vol. 41, 167–218. Burlington, VT: Academic Press.

Barrett, L.F., and J.A. Russell. 2014. *The Psychological Construction of Emotion*. New York: Guilford Publications.

Batson, C.D. 2008. Moral Masquerades: Experimental Exploration of the Nature of Moral Motivation. *Phenomenology and the Cognitive Sciences* 7: 51–66.

Bechara, A. 2005. Decision Making, Impulse Control and Loss of Willpower to Resist Drugs: A Neurocognitive Perspective. *Nature Neuroscience* 8: 1458–1463.

Bechara, A., D. Tranel, H. Damasio, and A.R. Damasio. 1996. Failure to Respond Autonomically to Anticipated Future Outcomes Following Damage to Prefrontal Cortex. *Cerebral Cortex* 6: 215–225.

Bechara, A., H. Damasio, D. Tranel, and A.R. Damasio. 1997. Deciding Advantageously Before Knowing the Advantageous Strategy. *Science* 275: 1293–1295.

Beer, J.S., O.P. John, D. Scabini, and R.T. Knight. 2006. Orbitofrontal Cortex and Social Behavior: Integrating Self-Monitoring and Emotion-Cognition Interactions. *Journal of Cognitive Neuroscience* 18: 871–879.

Björnsson, G., C. Strandberg, R.F. Olinder, J. Eriksson, and F. Björklund. 2015. Motivational Internalism: Contemporary Debates. In *Motivational Internalism*. New York: Oxford University Press.

Blair, R.J. 2007. The Amygdala and Ventromedial Prefrontal Cortex in Morality and Psychopathy. *Trends in Cognitive Sciences* 11: 387–392.

Blair, R.J., D.G.V. Mitchell, and K.S. Blair. 2005. *The Psychopath: Emotion and the Brain*. Malden, MA: Wiley-Blackwell.

Bok, H. 1996. Acting Without Choosing. *Noûs* 30 (2): 174–196.

Camp, E. 2009. Putting Thoughts to Work: Concepts, Systematicity, and Stimulus Independence. *Philosophy and Phenomenological Research* 78: 275–311.

Campbell, R. 2007. What Is Moral Judgment? *The Journal of Philosophy* 104: 321–349.

Campbell, R., and V. Kumar. 2012. Moral Reasoning on the Ground. *Ethics* 122 (2): 273–312.

Cholbi, M. 2006. Belief Attribution and the Falsification of Motive Internalism. *Philosophical Psychology* 19: 607–616.

Cima, M., F. Tonnaer, and M.D. Hauser. 2010. Psychopaths Know Right from Wrong but Don't Care. *Social Cognitive and Affective Neuroscience* 5: 59–67.

Copp, D. 2001. Realist-Expressivism: A Neglected Option for Moral Realism. *Social Philosophy and Policy* 18: 1–43.

Crockett, M.J. 2013. Models of Morality. *Trends in Cognitive Sciences* 17 (8): 363–366.

Damasio, A.R. 1994. *Descartes' Error*. New York: Putnam.

Dennett, D.C. 1995. *Darwin's Dangerous Idea*. New York: Simon and Schuster.

DePaul, M.R. 1991. The Highest Moral Knowledge and the Truth Behind Internalism. *The Southern Journal of Philosophy* 29 (Supplement): 137–160.

Diamond, C. 1982. Anything but Argument? *Philosophical Investigations* 5 (1): 23–41.

———. 1991. *The Realistic Spirit: Wittgenstein, Philosophy, and the Mind*. Cambridge, MA: The MIT Press.

Dretske, F. 1988. *Explaining Behavior: Reasons in a World of Causes*. Cambridge, MA: MIT Press.

Evans, G. 1982. In *The Varieties of Reference*, ed. J. McDowell. Oxford: Clarendon Press.

Fellows, L.K., and M.J. Farah. 2007. The Role of Ventromedial Prefrontal Cortex in Decision Making: Judgment under Uncertainty or Judgment Per Se? *Cerebral Cortex* 17: 2669–2674.
Frankena, W. 1958. Obligation and Motivation in Recent Moral Philosophy. In *Essays in Moral Philosophy*, ed. A.I. Melden, 40–81. Seattle: University of Washington Press.
Geach, P.T. 1956. Good and Evil. *Analysis* 17: 33–42.
Glenn, A.L., A. Raine, and R.A. Schug. 2009. The Neural Correlates of Moral Decision Making in Psychopathy. *Molecular Psychiatry* 14: 5–6.
Glimcher, P.W. 2009. Choice: Towards a Standard Back-Pocket Model. In *Neuroeconomics: Decision Making and the Brain*, ed. P.W. Glimcher, C. Camerer, R.A. Poldrack, and E. Fehr, 503–521. New York: Academic Press.
Glimcher, P.W., and E. Fehr. 2014. *Neuroeconomics: Decision Making and the Brain*. 2nd ed. New York: Academic Press.
Gordon, H.L., A.A. Baird, and A. End. 2004. Functional Differences Among Those High and Low on a Trait Measure of Psychopathy. *Biological Psychiatry* 56: 516–521.
Greene, J.D., R.B. Sommerville, L.E. Nystrom, J.M. Darley, and J.D. Cohen. 2001. An FMRI Investigation of Emotional Engagement in Moral Judgment. *Science* 293: 2105–2108.
Greene, J.D., L.E. Nystrom, A.D. Engell, J.M. Darley, and J.D. Cohen. 2004. The Neural Bases of Cognitive Conflict and Control in Moral Judgment. *Neuron* 44: 389–400.
Greene, J.D., S.A. Morelli, K. Lowenberg, L.E. Nystrom, and J.D. Cohen. 2008. Cognitive Load Selectively Interferes with Utilitarian Moral Judgment. *Cognition* 107: 1144–1154.
Griffiths, P.E., and A. Scarantino. 2009. Emotions in the Wild: The Situated Perspective on Emotion. In *Cambridge Handbook of Situated Cognition*, ed. P. Robbins and M. Aydede, 437–453. Cambridge: Cambridge University Press.
Haidt, J. 2001. The Emotional Dog and Its Rational Tail: A Social Intuitionist Approach to Moral Judgment. *Psychological Review* 108: 814–834.
Hare, T.A., C. Camerer, and A. Rangel. 2009. Self-Control in Decision-Making Involves Modulation of the VmPFC Valuation System. *Science* 324: 646–648.
Hilgard, E.R. 1980. The Trilogy of Mind: Cognition, Affection, and Conation. *Journal of the History of the Behavioral Sciences* 16: 107–117.
Holtzman, G. 2018. A Neuropsychological Challenge to the Sentimentalism/Rationalism Distinction. *Synthese* 195: 1873–1889.
Kable, J.W., and P.W. Glimcher. 2009. The Neurobiology of Decision: Consensus and Controversy. *Neuron* 63: 733–745.
Kahneman, D. 2011. *Thinking, Fast and Slow*. New York: Macmillan.
Kauppinen, A. 2015. Intuition and Belief in Moral Motivation. In *Motivational Internalism*, ed. G. Björnsson, F. Björklund, C. Strandberg, J. Eriksson, and R.F. Olinder, 237–259. New York: Oxford University Press.
Kelly, D., S. Stich, K.J. Haley, S.J. Eng, and D.M.T. Fessler. 2007. Harm, Affect, and the Moral/Conventional Distinction. *Mind & Language* 22: 117–131.
Kennett, J., and C. Fine. 2008. Internalism and the Evidence from Psychopaths and 'Acquired Sociopaths'. In *Moral Psychology*, ed. W. Sinnott-Armstrong, vol. 3, 173–190. Cambridge, MA: MIT Press.
Kiehl, K.A. 2006. A Cognitive Neuroscience Perspective on Psychopathy: Evidence for Paralimbic System Dysfunction. *Psychiatry Research* 142: 107–128.
Koenigs, M., D. Tranel, and A.R. Damasio. 2007a. The Lesion Method in Cognitive Neuroscience. In *Handbook of Psychophysiology*, 3rd ed. Cambridge University Press.
Koenigs, M., L. Young, R. Adolphs, D. Tranel, F. Cushman, M.D. Hauser, and A.R. Damasio. 2007b. Damage to the Prefrontal Cortex Increases Utilitarian Moral Judgements. *Nature* 446: 908–911.
Kriegel, U. 2012. Moral Motivation, Moral Phenomenology, and the Alief/Belief Distinction. *Australasian Journal of Philosophy* 90 (3): 469–486.
Kringelbach, M.L. 2005. The Human Orbitofrontal Cortex: Linking Reward to Hedonic Experience. *Nature Reviews Neuroscience* 6: 691–702.

Kringelbach, M.L., and K.C. Berridge. 2009. Towards a Functional Neuroanatomy of Pleasure and Happiness. *Trends in Cognitive Sciences* 13: 479–487.

Levy, D.J., and P.W. Glimcher. 2012. The Root of All Value: A Neural Common Currency for Choice. *Current Opinion in Neurobiology* 22 (6): 1027–1038.

Little, M.O. 1997. Virtue as Knowledge: Objections from the Philosophy of Mind. *Noûs* 31: 59–79.

McCormick, C., C.R. Rosenthal, T.D. Miller, and E.A. Maguire. 2016. Hippocampal Damage Increases Deontological Responses during Moral Decision Making. *Journal of Neuroscience* 36 (48): 12157–12167.

Mele, A.R. 2003. *Motivation and Agency*. New York: Oxford University Press.

Miller, E.K., and J.D. Cohen. 2001. An Integrative Theory of Prefrontal Cortex Function. *Annual Review of Neuroscience* 24: 167.

Moretto, G., E. Ladavas, F. Mattioli, and G. di Pellegrino. 2010. A Psychophysiological Investigation of Moral Judgment After Ventromedial Prefrontal Damage. *Journal of Cognitive Neuroscience*: 1888–1899.

Nichols, S. 2004. *Sentimental Rules: On the Natural Foundations of Moral Judgment*. Oxford: Oxford University Press.

Niv, Y., and P.R. Montague. 2009. Theoretical and Empirical Studies of Learning. In *Neuroeconomics: Decision Making and the Brain*, ed. P.W. Glimcher, C. Camerer, R.A. Poldrack, and E. Fehr, 367–387. New York: Academic Press.

Poldrack, R.A., and K. Foerde. 2008. Category Learning and the Memory Systems Debate. *Neuroscience and Biobehavioral Reviews* 32: 197–205.

Prinz, J.J. 2000. The Duality of Content. *Philosophical Studies* 100: 1–34.

———. 2002. *Furnishing the Mind: Concepts and Their Perceptual Basis*. Cambridge, MA: MIT Press.

———. 2004. *Gut Reactions: A Perceptual Theory of Emotion*. Oxford: Oxford University Press.

———. 2007. *The Emotional Construction of Morals*. Oxford: Oxford University Press.

———. 2015. An Empirical Case for Motivational Internalism. In *Motivational Internalism*, ed. G. Björnsson, F. Björklund, C. Strandberg, J. Eriksson, and R.F. Olinder, 61–84. New York: Oxford University Press.

Railton, P. 2014. The Affective Dog and Its Rational Tale: Intuition and Attunement. *Ethics* 124 (4): 813–859.

———. 2017. At the Core of Our Capacity to Act for a Reason: The Affective System and Evaluative Model-Based Learning and Control. *Emotion Review* 9 (4): 335–342.

Ridge, M. 2006. Ecumenical Expressivism: Finessing Frege. *Ethics* 116: 302–336.

Rolls, E.T. 2005. *Emotion Explained*. Oxford: Oxford University Press.

———. 2014. *Emotion and Decision Making Explained*. Oxford: Oxford University Press.

Roskies, A. 2003. Are Ethical Judgments Intrinsically Motivational? Lessons from 'Acquired Sociopathy. *Philosophical Psychology* 16: 51–66.

Russell, J.A. 2003. Core Affect and the Psychological Construction of Emotion. *Psychological Review* 110: 145–172.

Saver, J.L., and A.R. Damasio. 1991. Preserved Access and Processing of Social Knowledge in a Patient with Acquired Sociopathy Due to Ventromedial Frontal Damage. *Neuropsychologia* 29: 1241–1249.

Schacter, D.L., D.R. Addis, and R.L. Buckner. 2008. Episodic Simulation of Future Events: Concepts, Data, and Applications. *Annals of the New York Academy of Sciences* 1124: 39–60.

———. 2009. Constructive Memory and the Simulation of Future Events. In *The Cognitive Neurosciences*, ed. M.S. Gazzaniga, 4th ed., 751–762. Cambridge, MA: MIT Press.

Schnall, S., J. Haidt, G.L. Clore, and A. Jordan. 2008. Disgust as Embodied Moral Judgment. *Personality and Social Psychology Bulletin* 34: 1096–1109.

Schroeder, T. 2004. *Three Faces of Desire*. Oxford: Oxford University Press.

Seger, C.A. 2006. The Basal Ganglia in Human Learning. *The Neuroscientist* 12: 285–290.

Shafer-Landau, R. 2003. *Moral Realism: A Defence*. Oxford: Oxford University Press.

Shenhav, A., and J.D. Greene. 2014. Integrative Moral Judgment: Dissociating the Roles of the Amygdala and Ventromedial Prefrontal Cortex. *Journal of Neuroscience* 34 (13): 4741–4749.
Shohamy, D., C.E. Myers, J. Kalanithi, and M.A. Gluck. 2008. Basal Ganglia and Dopamine Contributions to Probabilistic Category Learning. *Neuroscience and Biobehavioral Reviews* 32: 219–236.
Shrager, Y., and L.R. Squire. 2009. Medial Temporal Lobe Function and Human Memory. In *The Cognitive Neurosciences*, ed. M.S. Gazzaniga, 4th ed., 675–690. Cambridge, MA: MIT Press.
Singer, P. 2005. Ethics and Intuitions. *The Journal of Ethics* 9 (3/4): 331–352.
Skyrms, B. 2010. *Signals: Evolution, Learning, and Information*. Oxford: Oxford University Press.
Smith, M. 1994. *The Moral Problem*. Malden, MA: Wiley-Blackwell.
Spreng, R.N., R.A. Mar, and A.S.N. Kim. 2009. The Common Neural Basis of Autobiographical Memory, Prospection, Navigation, Theory of Mind, and the Default Mode: A Quantitative Meta-Analysis. *Journal of Cognitive Neuroscience* 21 (3): 489–510.
Stanovich, K.E. 2009. Distinguishing the Reflective, Algorithmic, and Autonomous Minds: Is It Time for a Tri-Process Theory? In *In Two Minds: Dual Processes and Beyond*, ed. J.St.B.T. Evans and K. Frankish. Oxford: Oxford University Press.
Stevenson, C.L. 1937. The Emotive Meaning of Ethical Terms. *Mind* XLVI (181): 14–31.
Stocker, M. 1979. Desiring the Bad: An Essay in Moral Psychology. *The Journal of Philosophy* 76: 738–753.
Thagard, P. 2006. Desires Are Not Propositional Attitudes. *Dialogue: Canadian Philosophical Review/Revue Canadienne de Philosophie* 45: 151–156.
Thomson, J.J. 1985. The Trolley Problem. *The Yale Law Journal* 94: 1395–1415.
Tolhurst, W. 1995. Moral Experience and the Internalist Argument Against Moral Realism. *American Philosophical Quarterly* 32 (2): 187–194.
Tresan, J. 2006. De Dicto Internalist Cognitivism. *Noûs* 40: 143–165.
———. 2009. Metaethical Internalism: Another Neglected Distinction. *The Journal of Ethics* 13: 51–72.
Ugazio, G., C. Lamm, and T. Singer. 2012. The Role of Emotions for Moral Judgments Depends on the Type of Emotion and Moral Scenario. *Emotion* 12 (3): 579–590.
Valdesolo, P., and D. DeSteno. 2006. Manipulations of Emotional Context Shape Moral Judgment. *Psychological Science* 17: 476–477.
Wagner, D.D., M. Altman, R.G. Boswell, W.M. Kelley, and T.F. Heatherton. 2013. Self-Regulatory Depletion Enhances Neural Responses to Rewards and Impairs Top-Down Control. *Psychological Science* 24 (11): 2262–2271.
Wallis, J.D. 2007. Orbitofrontal Cortex and Its Contribution to Decision-Making. *Annual Review of Neuroscience* 30: 31–56.
Wallis, J.D., and S.W. Kennerley. 2010. Heterogeneous Reward Signals in Prefrontal Cortex. *Current Opinion in Neurobiology* 20 (2): 191–198.
Wheatley, T., and J. Haidt. 2005. Hypnotic Disgust Makes Moral Judgments More Severe. *Psychological Science* 16: 780–784.
Zarpentine, C. 2017. Moral Judgement, Agency and Affect: A Response to Gerrans and Kennett. *Mind* 126 (501): 233–257.
———. (n.d.). *Motivation Internalism and the Structure of Moral Psychology*. Manuscript

Chapter 6
The Reactive Roots of Retribution: Normative Implications of the Neuroscience of Punishment

Isaac Wiegman

Abstract Normative theories of punishment explain why punishment is morally justified, and these justifications usually appeal to two distinct considerations: retributive and consequentialist. In general, retributivist considerations are backward-looking and thus justify punishment in terms of "just deserts." By contrast, consequentialist considerations are forward-looking and thus justify punishment in terms of its good effects. This chapter contends that neuroscience should influence how we think about normative theories of punishment. First, it demonstrates how neuroscientific categories can distinguish the processes that guide human punishment decisions, especially decisions guided by retributive considerations. Second, this work in neuroscience supports an evolutionary debunking argument: Evolutionary explanations of these processes are a defeater for desert-based intuitions about punishment and the considerations that support it. More precisely, these intuitions are not good evidence for normative theories of punishment. This argument may have implications for a range of non-consequentialist principles and intuitions that extend beyond the domain of punishment. This chapter extends the evolutionary debunking argument to a broader set of intuitions, in particular, ones that are influenced by action-based behavior selection processes.

Keywords Evolutionary debunking · Model-based learning · Action and outcome · Retributivism · Consequentialism · Punishment

Desert is one of the central concepts of folk morality. Under this concept rewards, punishments, payments, debts, feelings of gratitude and resentment and much else are tabulated and dispensed. We punish, reward, and all the rest because of what a

---

I. Wiegman (✉)
Texas State University, San Marcos, TX, USA

G. S. Holtzman, E. Hildt (eds.), *Does Neuroscience Have Normative Implications?*, The International Library of Ethics, Law and Technology 22,
https://doi.org/10.1007/978-3-030-56134-5_6

person deserves given the quality of their actions or the content of their character. And with that, we can say also that desert is backward-looking. The desert-base has no immediate connection with the forward-looking concerns of consequentialist approaches to ethics, namely the consequences that attend rewards or punishments.[1] Instead, whether a person deserves punishment or reward depends more directly on the relation between what one has done in the past (or the content of their character) and what one receives in return. And so, desert has been a perennial bone of contention between the defenders and critics of consequentialism.

Nowhere is this more apparent than in the philosophy of punishment. There, philosophers have identified two classes of considerations in favor of punishment: retributive and consequentialist considerations. The distinctive feature of retributive considerations is their focus on whether punishment is deserved in light of past actions. Thus, the concept of desert captures the uniquely backward-looking character of retributive considerations. Henceforth, I will call a theory of punishment retributivist if it takes these "just deserts" considerations to be valid independently of the consequences of punishment. By contrast, the distinctive feature of consequentialist considerations is that they are forward-looking and concern only the good outcomes of punishment.[2]

In the ongoing debate between consequentialists and their critics, a recent development concerns the value representations that underlie human moral judgment. In particular, two (or more) distinct neural systems of learning and motivation operate on distinct value representations (Crockett 2013; Cushman 2013). One type of process, guides decision-making in a systematic, outcome-based manner. In other words, it places value on actions in accordance with the reward value of their outcomes. Accordingly, we can call these processes *outcome-based*, and it is notable that such processes appear to implement a fairly straightforward consequentialist decision procedure. Another type of process guides decision making in a heuristic manner, placing value on certain ways of acting, irrespective of whether their actual outcomes will be rewarded. Accordingly, we can call them *action-based processes*. Similarly, these processes correspond in some ways with non-consequentialist principles and intuitions (as I argue below).[3]

Given that action-based processes are simple and allegedly error prone, some philosophers think they are poor guides for moral theory (Greene 2008). Others have expressed skepticism about the implications of these and related findings for moral psychology (Berker 2009; Kahane 2012). Despite this resistance, many

[1] At least, so long as the consequences of an action are set in opposition "…to the circumstances or the intrinsic nature of the act or anything that happens before the act." (Sinnott-Armstrong 2003) For purposes of clarity, I will adopt this contrastive definition of consequences in what follows.

[2] Berman (2011) suggests that consequentialists might consider desert-outcomes in evaluating the total goodness in a world. This claim could only make sense on some alternative definition of consequence than the one I use here (cf. fn. 1). There is not space here to discuss the benefits of one definition or the other.

[3] Though see Kahane (2012).

remain optimistic that the science can shed light on the normative debate, either to debunk (Greene 2015) or vindicate anti-consequentialist intuitions (Kumar and Campbell 2012).

I think this optimism is warranted. Neuroscientific findings show that action-based processes contribute to retributive intuitions about punishment. Moreover, the evolved function of action-based processes undermines traditional appeals to retributive intuitions in normative theories of punishment (Wiegman 2017). Here, I extend this argument to encompass a wider range of intuitions, in particular, ones that are substantially influenced by individual learning processes. At the heart of this argument is the fact that action-based processes represent actions as valuable aside from their consequences, yet these actions are selected in part because of their favorable consequences. If this argument is correct, then the neuroscience of punishment, as well as value-guided learning, have normative implications for how we should think about the proper justification of punishment: there is substantially less support for the idea that it is intrinsically good to punish the wicked.

6.1 The Neuroscience of Reward/Valuation[4]

Neuroscience is helpful here because it uncovers fundamental behavioral, psychological, and neural distinctions between outcome-based processes and action-based processes. The former correspond with consequentialist reasoning and the latter correspond with reasoning about desert, or what I call "desert thinking" (perhaps among other non-consequentialist patterns of thought). In fact, the distinct patterns of behavior caused by outcome-based and action-based processes have been well understood for several decades. Nevertheless, work in neuroscience is beginning to underscore and vindicate this understanding by identifying the distinct neural causes of these patterns. So I will begin by describing the patterns of learning and behavior themselves, then briefly discuss the neuroscientific developments that shed light on them. The importance of these behavioral and neural distinctions is that action-based processes produce intuitions that are distinctively misleading concerning the moral value of punishment, or so I argue in the penultimate section.

6.1.1 Behavioral Phenomena

To start, consider classical conditioning, the most famous example of which is Pavlov's dogs. When presented with food, the dogs salivate, just like any other dog exposed to food. In this context, food is called an unconditioned stimulus (UCS),

[4]The discussion in this section owes a great deal to Balleine and O'Doherty (2010) and to Dayan and Berridge (2014). Many of the references in this section can be found in (Balleine and O'Doherty 2010).

because the salivation response to this stimulus is innate and thus is not a conditioned response to food. Dogs come relatively prepackaged with the ability to salivate in the presence of food. However, when the food is regularly presented with the ringing of a bell, the bell becomes a conditioned stimulus (CS) for salivation. In other words, the dogs learn to salivate at the sound of the bell, whether or not food is presented with it. Moreover, the salivation at the sound of the bell is difficult to unlearn. Even if food rewards are tied to the absence of salivation, dogs cannot unlearn salivation at the sound of the bell. At this point, two things are worth noticing about this form of conditioning. First, it is substantially inflexible since it is difficult to change the learned response to the conditioned stimulus. Second, this kind of learning is stimulus-bound rather than outcome oriented. Though salivation may improve the reward value of food (e.g. making it tastier and easier to digest), the dog does not learn this relationship (between response and outcome) through classical conditioning. Instead, the dog learns the connection between a stimulus (the bell) and an outcome (food presentation). As a result, we can call classical conditioning *stimulus-outcome learning*, or *S-O* learning (Balleine and O'Doherty 2010).

However, there is another form of learning that *does* associate a specific kind of response with the outcome of that response, and this form of learning takes place in instrumental conditioning paradigms. For instance, in rats, this kind of conditioning might involve delivering food pellets whenever the rat presses a bar. In this paradigm, the bar press is instrumental for food delivery. Early on in this kind of training, the rat is incredibly sensitive to the outcome of the bar press. In *contingency degradation* experiments (e.g., Balleine and Dickinson 1998), when the bar press no longer leads to delivery of food pellets (or if the pellets get delivered regardless of whether the bar is pressed), the rat will become less likely to press the bar. Similarly, in *outcome devaluation* experiments (Adams and Dickinson 1981), the rat might be injected with lithium chloride after eating the food pellets, so that it experiences nausea after eating them. Subsequently, it will become less likely to press the bar that delivers the pellets but no less likely to press other bars that deliver other consumables, such as sugar water, that have not been associated with nausea. What this suggests is that the rat learns to associate the action or *response* of pressing the bar with a specific *outcome*, the delivery of the food. We can call this form of learning, response-outcome learning or *R-O* learning (Balleine and O'Doherty 2010).

By contrast with S-O learning, R-O learning is highly flexible. From contingency degradation experiments, we see that the association only persists so long as the outcome is actually contingent on the response (whereas S-O associations persist even when the conditioned stimulus is no longer predictive of the unconditioned stimulus). Moreover, outcome devaluation experiments show that the rat's action of pressing the bar is also sensitive to how much the rat values the outcome. By contrast with classical conditioning, this kind of instrumental conditioning is not stimulus bound and is substantially flexible in relation to changing contingencies and valuations of outcomes.

One final form of learning needs to be considered here. This form of learning also occurs in instrumental conditioning paradigms, but only when the instrumental action (e.g. pressing the bar) has been over-trained in response to a conditioned stimulus or predictive cue, say a light flash that signals the availability of food via a lever press. Once the rat has been over-trained, it has acquired a kind of habit of pressing the bar when the light flashes. In effect, this kind of learning involves the association of some stimulus with a specific kind of response, so we can call this form of learning *S-R* learning (Balleine and O'Doherty 2010). By contrast with R-O learning, S-R learning is fairly inflexible in that it is relatively insensitive to changes in the contingency between response and outcome and insensitive to changes in the value of the outcome. Like S-O learning, behavioral responses to this form of learning are stimulus bound.

6.1.2 Distinct Processes of Learning and Motivation

Neuroscience has contributed to our understanding of these behavioral patterns by disentangling the distinct neural processes of learning and motivation that give rise to these patterns. These neural processes have distinct evolutionary histories and functions, and these differences in etiology will be important for the evolutionary debunking argument that follows. Also key is the difference in how these processes explain action. So, for each of these processes of learning and motivation, a main preoccupation will be this: when behaviors are performed under the influence of this kind of learning, why does the individual perform them? In the case of R-O learning, the behavioral evidence regarding outcome devaluation suggests a clear answer to this question. Individuals perform the instrumental behavior because they value its outcome, and will be less likely to perform it when they come to value that outcome less. So we can say that this kind of action is motivated by *outcome-based value representations*.[5]

Importantly, there is some evidence that R-O learning is governed by a specific form of reinforcement learning, called *model-based* learning, and this form of learning is implemented by neural substrates that are distinct from S-O and S-R learning. Algorithms that perform model-based learning have separate representations for states of the world, transitions between states, the values of each state, and a range of behavioral responses (see e.g. Dayan and Daw 2008). Moreover, this form of learning is model-based because for each behavioral response, the algorithm computes predictions of an outcome and its value using a model of the transitions between states and a model of the reward value of each state. The models are updated when actual rewards are received or when the system registers that state transitions have been modified. In the case of instrumental conditioning in rats, a prediction is made about the expected outcome of a bar-press, and once the outcome occurs, the expected outcome is compared to the actual outcome.[6] The difference

[5] I borrow this terms from Cushman (2013).

[6] The outcome is represented separately from its value in model-based learning algorithms (Dayan and Berridge 2014) and thus, there are two separate prediction-error signals.

between expected and actual state transition is the *state prediction-error*, which is used to update the rat's model of state-state transitions. By contrast, the difference between expected reward for a state and actual reward for a state is called *reward prediction-error*, which is used to update the rat's model of the reward function. These models are especially valuable for explaining R-O learning, in part because they show how actions can be motivated by a valued outcome. In effect, they explain how an action can be valuable to an agent because it brings about a specific outcome.[7]

Given that model-based learning requires distinct algorithms from other processes of learning and motivation (see e.g. Dayan and Daw 2008), it makes sense to ask whether model-based learning has a distinct neural substrate from other processes of learning and motivation. Studies in rats suggest that the prelimbic cortex and dorsomedial striatum are key components in model-free learning. Lesions in these areas make instrumental conditioning entirely habitual, even prior to over-training. Moreover, the prelimbic cortex is necessary for the acquisition of goal-directed behavior, suggesting that it plays some role in training the model of state-state transitions and of the reward function. In humans, there is some indication that the dorsolateral prefrontal cortex (dlPFC) plays a similar role in guiding model-based learning (Gläscher et al. 2010; Li et al. 2011), perhaps encoding state-prediction error.

It is one thing to learn about contingency relations between responses and outcomes, but it is another to know why an individual chooses to bring about that outcome. This is to ask how goal-directed behaviors are motivated. By contrast with S-R and S-O learning, goal-directed behaviors are not directly influenced by changes in appetitive motivational states, such as hunger or thirst. That is, R-O learning involves a rat experiencing the fact that food is more valuable when hungry *before* it adjusts behavior accordingly.

If we ask about the neural substrates of this evaluation process, it appears that in rats the basolateral amygdala plays a key role in integrating sensory (e.g. texture and smell) and affective inputs (e.g. how palatable or unpalatable a food is) to encode the value of a given outcome. When this area is damaged, outcome devaluation no longer results in the behavioral changes typically observed in R-O learning (Balleine et al. 2003). Data in humans is consistent with regions of the amygdala playing a similar role in encoding goal-value. There is also some indication that the ventromedial prefrontal cortex plays some role in encoding action-outcome values (Gläscher et al. 2009), though the role of this area is a source of ongoing contention.[8] Here is the point of characterizing the neural and behavioral distinctiveness of this and other processes of learning and motivation: these distinctions will be the basis for

[7]Though not moral agents, rats are clearly agents in a minimal sense, especially if they learn via model-based algorithms. In that case, they can decide between actions based on knowledge of contingencies between action and outcome. See e.g. Bermudez (2003), chapter 6.

[8]Another promising hypothesis is that it serves as a value comparator that decides between competing outputs of Pavlovian, model-free and model-based learning processes, perhaps together with their associated reward certainties.

debunking retributive intuitions about punishment without also debunking consequentialist ones.

The salient difference between these processes is suggested by the difference in R-O type behavior patterns, on the one hand, and S-O and S-R behavior patterns on the other: the former are driven by goal-value, whereas the latter are driven by two respective forms of value, which I will say are two different forms of *action-value*. By this, I mean roughly that an action that falls under these patterns is performed for its own sake, rather than for the sake of its expected outcome.[9] First, consider S-R learning. When instrumental behaviors are performed under the influence of this kind of learning, we can ask again, why does the individual perform these behaviors? By contrast with R-O learning, the answer does not refer directly to the value of the outcome. Remember that this form of learning is relatively insensitive to contingency degradation and outcome devaluation, meaning that the individual will continue performing this action even when it knows that a given outcome is less likely to occur and even when it values that outcome less. Instead, it appears that the individual performs these actions out of *habit* or because they are valuable as responses to certain stimuli. In other words, the action has *intrinsic* value. It is valued in itself as a response to a cue instead of being valued as instrumental for some outcome.

By contrast with R-O learning, S-R learning is well explained by *model-free* learning algorithms. As the name suggests, these learning algorithms do not include a model of state-state transitions or of reward values for each state. Instead, these algorithms determine the value of an action in a given state by averaging the value of its outcomes over time. Like model-based learning algorithms, model-free algorithms also use prediction-error signals, but unlike model-based systems, they use only *reward prediction-error* and not state prediction-error. In other words, this form of prediction-error is not based on predictions of state-state transitions (since model-free algorithms do not encode these transitions) but rather is based on predictions of the *reward* for an action in a specific state. As such, I will say that model-free algorithms operate on *action-based value representations*, since value is attached to specific actions rather than to specific outcomes.

Model-free learning processes appear to have distinct neural substrates from model-based processes: the dorsolateral striatum in rats or the putamen in primates. Lesions to this area result in R-O learning, even after overtraining. That is, habits never take hold, and as a result, behavior remains sensitive to contingency degradation and outcome devaluation (Yin et al. 2004).

Finally, consider S-O learning. In S-O learning, what is learned is not information about a certain kind of behavior response or its outcome. Instead, what is learned is an association between a stimulus and an outcome. In Pavlov's experiments, the result of this association is what we might call a *Pavlovian* response to

[9] By the word "action," I do not mean action in the highly intellectualized sense of "action for a reason" or actions that are performed because they are cast in a "favorable light" by their agents (McDowell 1982; Smith 1987). Rather, I mean the less intellectualized sense, common in the behavioral sciences, where action can just mean guided behavior.

the conditioned stimulus. In this specific case, the Pavlovian response is an innate response to food presentation, but more generally, we can think of Pavlovian responses as a diverse class of evolved responses to unconditioned stimuli that are innately rewarding (or punishing), such as food, sex, salt, water, temperature changes etc. In some cases, these responses are quite simple and reflexive, like salivating or making a gape face in response to an aversive taste or smell (Berridge 1996). In other cases, they could be flexible and extended over time, like moving toward an attractive object (e.g. food or a mate), or attacking an unfamiliar conspecific, perhaps even circumventing barriers to do so (cf. Seymour et al. 2007; Tindell et al. 2009). The point is just that these responses are evolved responses to certain kinds of unconditioned stimuli *in addition to stimuli that predict them*, namely conditioned stimuli.[10]

The details concerning Pavlovian responses are important for the evolutionary debunking argument below because a strong case can be made that certain retaliatory behaviors across several species are Pavlovian responses. For instance, work on intermale aggression in rats suggests an innate basis for territory defense (e.g. Blanchard and Blanchard 1984), and evolutionary models of resource competition explain why it would be (Wiegman 2019). Work on the frustration-aggression hypothesis also suggests that aggression is an innate response to frustration (where this is understood as the absence of an expected reward) and perhaps aversive events more broadly.[11]

In any case, S-O learning shows how Pavlovian responses can come under the control of cues that predict innately rewarding outcomes. *Pavlovian learning processes* then, are those that underpin this kind of learning. Similar to model-free and model-based learning algorithms, these processes also use prediction-error signals, which evaluate the predictive validity of conditioned stimuli for unconditioned stimuli.[12] In the case of model-based and model-free learning, prediction-error signals serve as a way of updating the value of a given outcome or action, but these signals play a different role here. Instead, they serve as a way of updating the predictive value of conditioned stimuli in relation to the unconditioned stimulus. The neural basis of Pavlovian learning processes appears to involve the ventral striatum, which probably encodes these prediction-error signals. Subregions of the ventral striatum also appear to implement Pavlovian responses.[13] The nucleus accumbens in rats influences Pavlovian approach and avoidance behaviors (Day et al. 2006).

---

[10] By saying that Pavlovian responses are evolved, I mean to say that they are heritable patterns of behavior that were selected for the increased fitness they confer on their possessors. This rules out various other kinds of explanations for such patterns of behavior, such as an explanation appealing to the development of these behavior patterns through reinforcement learning.

[11] For summaries of the relevant theory and evidence, see Berkowitz (1989, 2012).

[12] In fact, the discovery of dopamine neurons that increase firing rates prior to conditioned stimuli, was the first suggestion that that dopamine encodes prediction error signals (Schultz and Apicella 1992).

[13] Though I suspect that there is a wider range of Pavlovian responses than just approach and avoidance. In particular, there are different ways of approaching and avoiding candidate unconditioned

If we ask why individuals perform Pavlovian responses, we could give an evolutionary explanation of their behavior and a proximate explanation. The evolutionary explanation would be that the ancestors of these individuals were able to survive because they responded to a given unconditioned stimulus (e.g. pain) *and its indicators* in this way (e.g. flight or other kinds of avoidance). The proximate explanation would be that given the constitution of this particular organism, responses of this kind are intrinsically valuable as a response to innate rewards and punishments. As such, their value for the individual is not derived from their outcome. In other words, Pavlovian responses have action-value in that they are performed for their own sake, much like S-R-learned behaviors. We can see this by looking at analogs of contingency degradation and outcome devaluation experiments. Much like habits, Pavlovian responses to a conditioned stimulus will continue even when the conditioned stimulus no longer predicts the unconditioned stimulus and even when a particular instantiation of a unconditioned stimulus (e.g. food pellets) has been devalued (e.g. through satiation, Sheffield 1965). Moreover, like habitual behaviors and unlike outcome-oriented behaviors, Pavlovian responses are directly affected by changes in primary motivation (e.g. hunger and thirst, Dickinson and Balleine 1994). So, there is some reason to think that responses controlled by S-O and S-R learning are similar in kind.

6.1.3 Processes of Moral *Learning and Motivation*[14]

So what does all this have to do with punishment, desert, or even morality? These processes of learning and decision making are likely to be responsible for the distinct modes of moral evaluation captured by consequentialist and non-consequentialist approaches to morality. First, model-based learning has obvious affinities with (act) consequentialist decision procedures (Cushman 2013; cf. Greene 2008), which are based on the likelihood that an action will produce a given *outcome* as well as the *value* of that outcome. These elements of the decision procedure are clearly analogous to representations of state-state transitions and representations of reward functions (respectively) in model-based learning algorithms. So it is natural to predict that insofar as consequentialist decision procedures are implemented in the human brain, they will be implemented in brain regions associated with R-O learning. Several neuroscientific studies of moral judgment have found results consistent with this prediction (Greene et al. 2001, 2004; Greene 2009).

This raises the question of whether desert thinking might correspond to neural regions that are associated with S-O and S-R learning and thus action-value. While

stimuli. For instance, one candidate unconditioned stimulus for rats is a male conspecific with an unfamiliar smell, or an "intruder," and in some cases (e.g. in the presence of female rats or rat pups), the innate response to this unconditioned stimulus is to attempt to bite the back of the intruder.

[14] The discussion in this section owes a great deal to Cushman (2013) and Crockett (2013).

there are few direct neural ties between desert thinking and action-value, a connection is strongly suggested by the nature of desert thinking. As I pointed out above, desert thinking requires a tendency to give someone what they deserve, independently of whether this will result in a favorable outcome. Instead, desert thinking attaches value directly to actions that give people what they deserve. Model-free and Pavlovian processes provide a clear explanation of how this tendency could be implemented. Both processes use action-based value representations to motivate action in a specific context, and thus the value of actions is independent of their consequences, at least as the actions are represented by these systems.

Moreover, the value of the action is tied to a specific state or cue. It is not just that attacking an intruder or habitually pressing a bar is *always* represented as valuable, but rather, it is represented as valuable in a specific context or in response to a specific elicitor. As a result, these processes can explain in part why it matters to us to give people what they deserve. At the very least, it is less obvious how model-based learning could explain the outcome independence of desert-based judgments. If we only know about three forms of learning and one of them (model-based learning) doesn't explain the action in question, then we have good reason to suppose that the action is explained by one of the other two. Given that we have some framework for understanding the state of the world in terms of what others have done (we obviously do), model-free and Pavlovian processes can explain how action-based value representations attach to various ways of responding to such states. The result would be that certain actions (e.g. revenge) are represented as valuable responses to what others have done. While this is not yet close to the kind of empirical vindication that we might want, psychologists have already begun showing how otherwise puzzling patterns of moral judgments (such as the differential influence of outcome-based processes on moral decisions as manifested, for example, in the Trolley problem) can be explained by the interactions of these systems (Crockett 2013; Cushman 2013). Thus, it appears that this multi-system framework for moral judgment is promising for explaining a range of anti-consequentialist patterns of decision-making.

Of course, this leaves room for doubt. More importantly, the explanation above is unlikely to be the entire explanation for desert thinking generally.[15] It leaves open many interesting questions: Why do we care about some things that people do rather than others? Why do these processes identify certain types of actions (e.g. murder and theft rather than hunting and foraging) as basis for dispensing punishments and rewards? How do these processes determine the range of responses that we consciously judge to be appropriate in response to certain types of actions? These

[15] This is because desert thinking does not just involve habits and impulsive responses, but rather, also includes patterns of thought and judgment. This explanatory gap is not unbridgeable by any means. There is a well-documented trend toward motivated reasoning in humans (e.g. Nisbett and Wilson 1977). In particular, this involves a tendency to rationalize one's behavior post hoc in order to promote consistency between actions and conscious deliberation. Greene and Haidt have suggested that these processes play an important role in conscious moral reasoning (Greene 2008; Greene and Haidt 2002; Haidt 2001).

questions and many others can only be answered by a complex story that integrates the interaction and development of these processes with dynamic social experiences. That is a monumental task.

For these and many other reasons, it would be extremely difficult to pull together evidence tying desert thinking generally to the specific brain regions involved in Pavlovian and model-free learning. Desert is a very broad phenomenon, with many and diverse behavioral manifestations. Some philosophers have even suggested that there are different kinds of desert claims, each of which has an independent justification (e.g. Sher 1987). These are just a few of the reasons why I favor a "divide and conquer" strategy for explaining desert thinking. This approach identifies specific domains of desert thinking and attempts to tie them to specific processes of learning and motivation. Desert thinking in the domain of punishment is an ideal place to start.

6.2 Punishment and Action-Based Value Representations

As I suggested above, punishment decisions may be an outgrowth of Pavlovian responses and processes with a long evolutionary history. The path to this conclusion begins with the philosophy of punishment, which reveals a key distinction among motives for punishment.

6.2.1 Punishment in Philosophy

Perhaps the central philosophical concern with punishment is in its justification. Punishment usually requires the imposition of hard treatment such as inflicting harm, pain or suffering, withholding concern or goodwill, or perhaps suspending the rights of the transgressor. Such activities are seldom warranted without strong justification. So what are the considerations that favor or justify punishment?[16]

---

[16] Notice that these positive considerations are distinct from the considerations that favor non-punishment (e.g. innocence) or that constrain the severity of punishment (e.g. seriousness of the offense). This is important because I am offering a criticism of certain core commitments of retributivism involving positive considerations, but some philosophers define retributivism so as to include the negative considerations as well. I believe that intuitions concerning negative retributive considerations have an independent psychological basis and are relatively independent of the argument I give against positive retributive considerations.

Another reason to focus on positive considerations is that the logic of mitigating factors only makes sense against a backdrop of certain positive considerations for punishment. Mitigating punishment for (or excusing) accidental harms only makes sense if we have some reason to think that non-accidental harms warrant punishment. If the ultimate justification for such practices is consequentialist in nature (e.g. Rawls 1955), then we can likewise explain the excusing conditions in

As pointed out in the introduction, philosophers have canvassed two types of considerations in favor of punishment: retributive considerations and consequential considerations. The distinctive feature of retributive considerations is their focus on whether punishment is deserved in light of past actions. Thus, the concept of desert captures the uniquely backward-looking character of retributive considerations. By contrast, the distinctive feature of consequentialist considerations is that they are forward-looking and concern the good outcomes of punishment. As such, Nadelhoffer et al. (2013) include all of the following as consequentialist considerations for punishment:

1. General deterrence – i.e. punishing in order to deter other would-be offenders from committing similar offences.
2. Incapacitation – i.e. punishing in order to prevent the offender from committing similar crimes while he is being detained and/or treated.
3. Rehabilitation and moral education – i.e. punishing in order to rehabilitate or re-educate the offender…
4. Catharsis – i.e. punishing in order to give victims and society more generally a healthy emotional release.
5. Norm reinforcement – i.e. punishing in order to highlight and reassert the importance of social values and norms.
6. Quelling revenge – i.e. punishing in order to keep the original or third parties from starting a blood feud. (Nadelhoffer et al. 2013, p. 237)[17]

Each of these considerations can be understood as reflecting concern only with the outcomes of punishment, and we can contrast these considerations with considerations of desert. When one's concern is with giving an offender their "just deserts," the point is to match the punishment with crime that the person has committed. On this view, if we ask "why punish this person?," the answer will be likely to include considerations like the following:

1. Severity of harm done
2. Moral quality of behavior (how offensive or admirable it is)
3. Good/bad intentions behind action
4. Blameworthiness/praiseworthiness
5. Negligence/responsibleness of behavior (adapted from Carlsmith and Darley 2008, pp. 123–4)

---

terms of the consequences of accepting certain excuses as a matter of policy: e.g. it would cause less harm in certain cases without undermining general deterrence.

[17] One might also include under this heading the value of enforcing threats that one has a moral right to make (see e.g. Quinn 1985). Nadelhoffer et al.'s (2013) list includes the following: "4. Communication – i.e. punishing in order to communicate or express disapproval of an action." Nevertheless, this clearly falls outside of consequentialist considerations as I understand them here, because one clearly cannot specify the value of communication without referring to past actions of the offender. Cf. fns. 106 and 107.

What is important here is that each of these items concerns the past, and none has any direct relevance to assessing the outcome of punishment.[18] So we can say that if punishment is justified by these considerations, punishment is backward-looking.

This brings us to the defining feature of retributivism in particular and desert thinking more generally. Given that someone thinks it good or valuable to give people what they deserve, including punishment, it follows that they also take punishment to have value that is not derived from its consequences. We can say equivalently that they take punishment to have *intrinsic* value.

So what reason is there to believe that punishment has intrinsic value? One of the main justifications is based on moral intuitions, or non-inferential moral judgements. Michael Moore gives voice to this justification:

> My own mode of justifying retributivism has tried to do this…I take seriously the sorts of particular moral judgments that…thought experiments call forth in me and in most people I know:…for example, Dostoevsky's Russian nobleman in *The Brothers Karamazov*… Question: should…[the] offender be punished, even though no other social good will thereby be achieved? The retributivist's 'yes' runs deep for most people. (Moore 2010, p. 163)

From the context, it is obvious that Moore takes these kinds of judgments to be non-inferential. For, if there were some inference on which they are based, the justification of punishment would depend on the premises of such an inference rather than on the judgment itself. But where does the "retributivist's 'yes'" come from? How do these moral judgments arise if not by inference from some epistemically prior belief?

6.2.2 *Punishment Is Not Model-Based*

The most plausible answer is that they arise out of processes that place value directly on certain ways of acting, rather than on the outcomes of those actions. The central direction of work in social psychology and behavioral economics confirms this suspicion: punishment decisions in general are *not* aimed at securing a range of salient *outcomes*. Instead, they are the product of processes that place value on *actions*, specifically ones that pay the transgressor back for their transgression. There are two key predictions that one can make about people's punishment decisions based on the putative contribution of action-based or outcome-based value representations. First, if punishment decisions are influenced by action-based value representations and if these value representations correspond with desert thinking, then the severity of punishment will change in response to various factors that influence

[18] While most of them obviously have indirect relevance to outcomes, these can be dissociated. The motive to punish those with bad intentions might be thought to deter others who have bad intentions and a general awareness that people aim to punish bad intentions. Nevertheless, in light of the evidence below, this is clearly not the line of reasoning that motivates people to punish those with bad intentions.

desert. In general, people will judge that more severe punishments are appropriate when the offender intentionally committed a serious moral offence that caused a great deal of harm and for which the offender is morally responsible (e.g. their judgment was not impaired). Second, if punishment decisions are influenced by outcome-based value representations, then the severity of punishment will change in response to changes in the consequences of punishment (e.g. its deterrent value and its foreseeable side effects). For instance, probability of detection determines the severity of punishment required to achieve a certain level of deterrence. If a crime is more difficult to detect, then it requires more severe punishment to reduce the expected utility of the crime. So if someone's punishment judgments are motivated by the aim of deterrence, then the severity of punishment should increase as the probability of detection decreases.

The latter prediction has not born out. Over the course of nine studies, Baron and Ritov (2009) asked both judges and internet participants to assign penalties for various crimes and also had them rate the seriousness of the crime and their anger at the crime (in some of the experiments). In almost every case, the seriousness of the crime or anger in response to it were much better predictors of the severity of punishment than the probability of detection for the crime. Only a few participants' severity assignments tracked the probability of detection. Moreover, only when probability of detection was highly salient did it influence severity of punishment in the direction predicted by consequentialist motives. Contrary to their predictions, Baron and Ritov found that even when participants took on a policy-making perspective as opposed to making judgments about a specific violation (e.g. "This item is about how future offenses should be penalized" as opposed to "about an offense already committed.", p. 572), lower probability of detection did not lead to increased severity of punishment.

Carlsmith (2008) explored the influence on punishment of a wider range of consequentialist considerations (e.g. "the publicity of the crime and subsequent punishment, the frequency of the crime, the likelihood of similar crimes in the future, the likelihood of detecting the crime, and the likelihood of catching the perpetrator." Carlsmith 2008, p.124) and retributive considerations (e.g. "the severity of the harm, the moral offensiveness of the behavior, the intent behind the action, the blameworthiness of the offender, and whether or not the offender was acting in a responsible manner." Carlsmith 2008, pp. 123–124). As a partial replication of previous work with colleagues (Carlsmith et al. 2002; Darley et al. 2000), he found that consequentialist considerations were poor predictors of people's actual sentencing decisions, whereas retributive considerations were strong predictors. Even though people's stated motives for punishment usually focused on deterrence, these reports did not correlate at all with people's actual decisions.

In another study, Carlsmith (2006) investigated people's decision making process by allowing participants to selectively and sequentially access different information about a given criminal offense, probing for severity of punishment ("using a scale ranging from 1 not at all severe to 7 extremely severe," p. 444) and confidence in sentence after each selection. Some of the information concerned retributive considerations (e.g., "Magnitude of harm", "Perpetrator intent", "Extenuating circumstances"). Other information had more to do with consequentialist considerations (e.g., "Likelihood of violence", "Prior record", "Self-control", "General frequency", "Detection rate", "Publicity"). The key result was that people frequently chose to access information relevant to retribution prior to accessing information relevant to deterrence. Moreover, Carlsmith reported that "…retribution information improved confidence more than did incapacitation information," that is, information about punishment outcomes (Carlsmith 2006, p. 446).

Importantly for the discussion below, some of these studies (including Baron and Ritov 2009) indicate that anger is connected to retributive judgments. For instance, Carlsmith et al. (2002) found that in response to vignettes about punishment, moral outrage ratings were a strong predictor of punishment and mediated the influence of retributive considerations (i.e. culpability and seriousness of offense) on those judgments.

One constraint in generalizing the results of these studies is that the punishment preferences recorded in these studies were entirely hypothetical. Moreover, there is often a large gap between stated preferences and actual behavior. One might wonder whether punishment becomes more consequentialist where the rubber meets the road. Work in behavioral economics is perhaps the best place to test such a hypothesis, since these studies often have people interacting in real time and competing for real monetary rewards. Nevertheless, even in studies such as these, punishment decisions do not appear to be outcome oriented.

Consider for instance *public goods games*. In these economic games, a group of several people are each given a starting endowment of money that they can choose to invest together in a "group project." Whereas each individual gets to determine how much to invest in the project, the return on the investment is distributed evenly to everyone in the group. So there are strong incentives not to invest in the project. If everyone but me invests, then I will profit just as much as everyone else while also getting to keep my initial endowment. In some versions of this game (e.g. Fehr and Gächter 2002), each round of the game is an anonymous, one-shot interaction facilitated by computers. That is, if I were to play for eight rounds, I would play with different participants on each round, and I would not have any identifying information about any of the other participants. Nevertheless, at the end of each round, I would receive information about how much the other players invested in the project.

Given the incentives to free ride, it is no surprise that investment in group projects drops the more rounds are played. Nevertheless, in other versions of the game, each round ends with an opportunity to punish other participants by paying a small cost. For instance, if during one round, I discover that another participant invested only $2 in the group project, when everyone else invested $5, then I could pay $1 so that this participant loses $3 (or $2 so they would lose $6). In this version of the game, punishment is fairly frequent and usually has the effect of dramatically increasing investments in the group project. Nevertheless, it is clear that individuals are not punishing in order to increase their own returns. If I pay to punish another player on round one, I know that I will not interact with that player again, so it seems unlikely that I am punishing them to increase my future profits.

Perhaps then participants are punishing with the aim of increasing other people's profits. This motive would make sense of some cases of punishment in the public goods game, but it does not make sense of punishments that occur on the last round of the game. It is doubtful that anyone expects to secure a good outcome from punishments at this stage of the game. Rather it is more plausible that participants punish only because free riders deserve it, though perhaps the more secure conclusion is just that punishment is not driven by outcome-based value representations. If it were, then we would predict greater sensitivity to a wide range of relevant goals (e.g. deterrence, maximizing monetary rewards, etc.). A shakier, but nevertheless tempting conclusion is that these punishment behaviors are driven by action-based value representations. If actions of punishing offenders and free riders are valued in themselves, this provides a compelling explanation of why participants in these various studies would, for instance, punish on the final round of the public goods game or assign more severe punishments that are unnecessary for greater deterrence. This conclusion is strengthened by various lines of evidence that punishment is at root a Pavlovian response and is therefore guided by action-based value representations.

6.2.3 Pavlovian Punishment

As will be outlined below, there are five lines of mutually supporting evidence that retributive punishment is a Pavlovian response at root. These same lines of evidence suggest that retributive punishment is an innate adaptation. While no one of these strands of evidence is strong on its own, I find their collective strength compelling.

6.2.3.1 Affective Influences

As mentioned before, it is common for participants' ratings of their outrage at an offense to track the severity of punishments they assign (see also Nelissen and Zeelenberg 2009). Importantly, none of the studies so far mentioned manipulated anger to measure its effects on punishment judgments. However, others (Ask and Pina 2011; Goldberg et al. 1999; Lerner et al. 1998) have done just that. In one of these experiments (Lerner et al. 1998), one set of participants, the anger induction group, watch a video depicting a bully and an accomplice who assault and humiliate a teenager. Another set of participants, the control group, watched a video of abstract colors and shapes with negligible emotional content. Afterwards, participants read vignettes describing hypothetical harms and rated the degree to which the perpetrators ought to be punished. The punishment ratings of the anger induction group were higher than controls, demonstrating that incidental anger influences judgments about punishment. This is important for two reasons. First, anger is widely understood as an irruptive motivational state that motivates shortsighted payback behaviors, interrupting or overriding "better judgments" to do so, suggesting that it is not guided by goal-based value representations. Second, anger is among a small number of putatively "basic" emotions, that are thought to be innate adaptations for solving recurrent problems in evolutionary history (Ekman and Cordaro 2011). Pavlovian responses generally are thought to have the same kind of adaptive function.

6.2.3.2 Universality

There are strong indications that retributive punishments are universal. For instance, Henrich et al. (2006) found evidence of costly punishment of norm violators in an economic game across a wide range of cultures, including modern day hunter-gatherers, pastoralists and horticulturists. Similarly, in Daly and Wilson's (1988) review of the ethnographic record, they found clear evidence of retributive norms and behaviors in over 95% of cultures. Moreover, in many cultures, norms of revenge function to limit or restrain revenge (cf. Daly and Wilson 1988, Chapter 10). As Robert Frank suggests, "We may safely presume that, where a cultural norm attempts to restrain a given behavior, people left to their own devices would tend to do even more of it." (Frank 1988, p. 39). Thus, together with facts about cultural restraints on retribution, the universality of this trait makes it unlikely to be purely a product of cultural or individual learning. Rather, it is more likely that the trait is innate or perhaps buffered against environmental variability.

6.2.3.3 Early Development

This hypothesis is further supported by the early development of retributive behaviors. Children younger than 2 years of age will "punish" characters in puppet morality plays who hinder the efforts of other characters (Hamlin et al. 2011). Moreover, infants as young as 5 months will show a preference for characters who harm hinderers over characters who helped them (Hamlin 2014). The latter effect shows a very early, if primitive, grasp of the thought that punishing a bad person is a good thing. Given the low likelihood that many 5 month old infants have observed acts of punishment, much less seen others approve of such acts, it is plausible that some of the relevant mechanisms are innate.

6.2.3.4 Neural and Genetic Underpinnings[19]

Finally, punishment behaviors in humans appear to have neural and genetic underpinnings that are shared with revenge behaviors. Strobel et al. (2011) conducted fMRI scans while participants participated in or observed a dictator game, in which another player (actually a computer program) received a monetary endowment and decided how much of it to split with another player (either with the participant or with a third party). Participants then had the option to pay money in order to punish the dictator. Players punished the dictator when the split was unfair whether they were on the receiving end of the unfair split or whether they merely observed someone else on the receiving end. Strobel and colleagues observed activation in the nucleus accumbens during punishment, as well as in other brain regions associated with reward (replicating results from de Quervain et al. 2004), but more interesting for our purposes, they also analyzed the effects of different alleles for a gene that influences dopamine turnover. In a regression model, variation in alleles for this gene accurately predicted higher levels of activation during punishment in the nucleus accumbens, and did so whether the punishment occurred in the observer or receiver condition. Given the plausible assumption that revenge played a role in the receiver condition, it is likely that impartial forms of punishment (as in the observer case) are motivated by some of the same neural structures that motivate revenge. This is significant because, as I argue above, revenge is itself likely to be a Pavlovian response.

6.2.3.5 Evolutionary Explanation

In fact, we should expect revenge and retribution both to be influenced by Pavlovian processes, because evolutionary models of social interactions suggest that both are adaptive strategies, if for slightly different reasons. Revenge appears to be adaptive because it deters other individuals from harming oneself or one's kin. The thought is that when one gets revenge, one "makes an example" of the target. As a result, both the target and the audience of revenge will be less likely to offend in the future (Clutton-Brock and Parker 1995; Frank 1988; McCullough et al. 2012).

[19] There is a wealth of research on the neural underpinnings of punishment decision in humans, more than I could review in the space allotted. Interestingly, many of the brain regions implicated in this research include those listed above, in connection with model-based, model-free, and Pavlovian action selection processes. As with almost any decision that humans make, punishment decisions appear to be a product of all of these action selection mechanisms working in tandem. Nevertheless, it is important that the bulk of this research recognizes that punishment is based on "just deserts" considerations (rather than consequentialist considerations), and so the role of regions like the dlPFC is not likely weighing the outcomes of punishment (see e.g. Buckholtz et al. 2015). Whatever the role of the dlPFC is in punishment decisions, it is an independently evident point that the punishment decisions it arbitrates are not guided entirely by outcome-based value representations.

Retributive punishment of norm violators confers two advantages on large cultural groups that trickle down to the individuals in those groups. First, it stabilizes cooperation within larger groups by deterring free-riding. This probably afforded our ancestors advantageous forms of specialization and niche construction (e.g. Gintis et al. 2008). Second, it stabilizes cultural variation between large cultural groups despite migration (e.g. Richerson and Boyd 2005). The latter role allows large cultural groups to retain cultural variants that help them survive in their local environments. By contrast, if immigrants from another environment (e.g. with different food sources and different parasites) brought their traditions and food preparation customs with them to a new group and if those customs were to spread within the new group, the new groups lose the customs that help them survive in the local environment or they could more easily gain maladaptive customs. What works in one ecological niche may not work in another.

6.2.4 Summary on the Value Representations that Guide Punishment

Pulling all this together, neuroscientific categories help to identify the value representations that underlie punishment. There is little evidence that retributive judgments and actions are driven by outcome-based value representations. Instead, they bear the marks of an innate, adaptive behavioral response, akin to the Pavlovian responses and learning processes that underpin motives like hunger and thirst (e.g. Berridge 1996; Dayan and Berridge 2014). As responses to certain kinds of stimuli (e.g. unconditioned stimuli), these responses are represented as valuable in themselves and their value is not mitigated by information about their effects, even if an individual has access to such information.

6.3 An Evolutionary (or Rather, Selection-Based) Debunking Argument

Not only do neuroscientific categories help identify the value representations behind punishment, they also shed light on how these values arise. As will be discussed below, actions of punishment are represented as valuable independently of their outcomes, yet these actions are valued in large part because of their past outcomes. This contrast between representation and etiology results in an undercutting defeater for retributive intuitions as support for retributive theories (such as the "retributivist's 'yes'" discussed above). To see this, consider again how actions acquire action-value through model-free or Pavlovian processes.

First, consider model-free learning processes and the acquisition of habits. Given that habits are acquired through processes of learning, their mode of acquisition is through individual learning via reward (or punishment). Model-free learning algorithms establish habits because they were reliably rewarded in the past (e.g. due to overtraining). In other words, habitual actions are selected by learning algorithms because of their past outcomes (though not because of the organisms' expectations concerning immediate outcomes).

Something similar can be said of Pavlovian responses, except with regard to their acquisition by species rather than individual organisms. For instance, salivation in the presence of food and food-predictors was acquired over evolutionary time because of the adaptive benefits it bestows on the organisms that possess it. At a small risk of oversimplification, organisms that salivated appropriately were better able to digest their food and pass their genes along to subsequent generations. In other words, Pavlovian responses were *selected* by evolutionary processes because of their outcomes. If the adaptive story concerning punishment is correct, then punishment (qua norm enforcement) exists in humans today for similar reasons, except that the good outcome accrued to cultural groups first and to individuals as a result.

Either way, we have a selection-based explanation of action, where actions are selected by a process because of their good outcomes. In the case of habits, learning algorithms select the action because of its past rewards, and in the case of Pavlovian responses, evolutionary processes select the action because of its fitness-enhancing effects. Nevertheless, (as I argued above) these actions are represented as intrinsically valuable, or valuable independently of their consequences.

Moreover, we are now in a position to see that these actions appear intrinsically valuable to an individual because it is part of the function of both of these selection processes to cause agents to act without access to information about the likely effects of the action. In the case of evolution, this is because organisms need to be prepared for some situations (e.g. encounters with predators and hostile conspecifics) without knowing anything about the likely effects of their actions. In the case of model-free learning algorithms, it frees up valuable cognitive resources to act without calculation of the outcomes of so acting.

6.3.1 Intrinsic Value Debunked

If I understand him correctly, Moore's argument above (in the section entitled "Punishment in Philosophy") is that intuitions about the intrinsic value of punishment are at least prima facie evidence in support of retributive theories of punishment. Yet this argument only stands so long as the evidential value of the intuitions is not overturned or defeated by other evidence. We are now in a position to see how the evidential value of these intuitions is overturned by the facts about Pavlovian and model-free processes.

Suppose I am right that punishment intuitions are underpinned by these processes. Whereas actions of punishment are represented as intrinsically valuable, the processes that give rise to them are not good indicators of intrinsic value. To reiterate, intrinsic value of actions is value that is separate from effects for which the action is instrumental. Such value cannot be reliably attributed to an action by a process (evolutionary or developmental) that selects actions as instrumental for greater fitness or reward or whatever. So, if I am right that punishment intuitions, judgments and decisions are guided by Pavlovian processes, selected for their fitness-enhancing effects, then these intuitions could not possibly be reliable indicators that punishment has intrinsic value.

Another way of putting this point is that evidence concerning the etiology of Pavlovian responses (or model-free processes) severs the evidential link between punishment intuitions and the conclusions that they putatively support (as with an *undercutting defeater*).[20] By comparison, one might ordinarily trust the testimony of an expert witness, but if evidence is presented that the expert witness has a conflict of interest, the evidential value of her testimony is compromised. Evidence concerning a conflict of interest severs (or at least compromises) the evidential link between the expert's testimony and the conclusions that the testimony would otherwise support. Selection-based considerations play this exact role. They show that retributive intuitions arise because of behavior selection processes that select on the basis of outcomes. These intuitions are like testimony that actions have intrinsic value (i.e., value that is independent of outcome), but the processes that produce these intuitions have something like a conflict of interest: they originated in the service of outcomes. Given the conceptual division between outcome-based value and intrinsic value, a process that selects actions in service of outcomes (e.g., ones that increase chances of survival and reproduction) cannot reliably indicate that the action has intrinsic value. So, if the Pavlovian processes that give rise to punishment intuitions are ultimately explained by the outcome-value of punishment (i.e., its survival value), then it cannot reliably indicate that punishment has intrinsic value. So we are in a position to doubt whether there is any evidential link between retributive intuitions and conclusions about the intrinsic value of punishment. In other words, it would appear to be a coincidence if Pavlovian processes selected actions that really were intrinsically valuable.[21]

By contrast, model-based processes clearly are sensitive to whether an action is instrumentally valuable, and it represents actions as such. This is because model-based processes motivate an action (e.g., bar presses) only if it is instrumental for some valued outcome (e.g., a food item that the individual values). As a result, there is no severing here of the connection between the outputs of model-based processes and the conclusions drawn from them. The conclusions we draw are that this or that action is valuable because of its good outcomes, and the process that gives rise to these judgments is exquisitely sensitive to the value and likelihood of outcomes, as demonstrated in outcome devaluation and contingency degradation experiments.

---

[20] Another way to defeat a piece of evidence is to show that other evidence overrides it, as with a *rebutting defeater* (Pollock 1987).

[21] I draw out this argument in greater detail in Wiegman (2017).

6.3.2 *Thesis 2: Evolutionary Explanation Is a Defeater*

Thus, evolutionary explanations of retributive intuitions serve as a defeater. These intuitions can no longer be used as evidence for retributive theories, and something similar could be said of any other desert-based intuition for which we have a selection-based explanation. That is, if we can pin the development of desert-based intuitions on model-free or Pavlovian processes of learning and motivation, then we have a ready defeater for their evidential value in deciding between consequentialist and non-consequentialist theories of moral value.[22]

6.4 Conclusion

The path to this conclusion has been long and the terrain varied, so I conclude by retracing our steps. I began by pointing out that the normative relevance of desert to punishment is a perennial point of contention between consequentialist and non-consequentialist theories of value. To say that certain actions are justified because of considerations of desert is to say that the actions are justified even if they do not produce good consequences. This is to say that these actions have intrinsic moral value, or moral value independent of their consequences. This applies mutatis mutandis to the philosophy of punishment. To say, as the retributivist does, that punishments are justified because punishees deserve them is to say that actions of punishment have moral value aside from their consequences.

Moreover, some of the strongest pieces of evidence in favor of retributivism are our intuitions about the appropriateness of punishment in various cases. Yet evidence from the psychology and neuroscience of reward and motivation suggests at most three possible origins for these intuitions: model-based, model-free, and Pavlovian processes. When we look at the psychology and neuroscience of punishment, Pavlovian processes are the most likely suspect. We also know that these processes represent actions as valuable in response to certain stimuli and independently of the consequences of so responding. Yet, this valuation is deceptive. For the evolutionary processes that gave rise to Pavlovian responses (and the model-free processes that give rise to habits) select actions on the basis of their outcomes. The presentation *as of* "actions with intrinsic value" is thus elusory. In the end, we have less evidence than we originally thought for the claim that punishment has intrinsic value. This is a normative conclusion that is substantially informed by neuroscience.

It might seem that the consequence is that we must do away with desert thinking wholesale, at least in the domain of punishment. Nevertheless, this is the wrong

[22] Of course, an intuition could be explained by both outcome-based processes and action-based processes, but if the intuition attributes intrinsic value to an action, this aspect of the intuition is most likely caused or conditioned by action-based processes.

conclusion to draw if desert thinking itself can be given a consequentialist rationale. The challenge to desert thinking here presented is to question whether it has any basis *aside from its consequences*. But many philosophers have suggested that desert thinking has good consequences. For instance, social organization is possible in large part because we have a practice of seeking out and punishing those who deserve it. Of course, the value of this practice depends in large part on how it is implemented. In fact, practices of punishment in America lead to disproportionate and large-scale incarceration of minorities (Alexander 2012), and these effects are certainly among those that should constrain or mitigate our practices of assigning punishments. If we can step away from our most basic retributive inclinations, then perhaps we can have the clarity with which to reform these egregious injustices.

References

Adams, C.D., and A. Dickinson. 1981. Instrumental Responding Following Reinforcer Devaluation. *The Quarterly Journal of Experimental Psychology Section B Quarterly Journal of Experimental Psychology* 33 (2): 109–121. https://doi.org/10.1080/14640748108400816.

Alexander, M. 2012. *The New Jim Crow: Mass Incarceration in the Age of Colorblindness*. New York: New Press.

Ask, K., and A. Pina. 2011. On Being Angry and Punitive: How Anger Alters Perception of Criminal Intent. *Social Psychological and Personality Science* 2 (5): 494–499. https://doi.org/10.1177/1948550611398415.

Balleine, B.W., and A. Dickinson. 1998. Goal-Directed Instrumental Action: Contingency and Incentive Learning and Their Cortical Substrates. *Neuropharmacology* 37: 407–419.

Balleine, B.W., and J.P. O'Doherty. 2010. Human and Rodent Homologies in Action Control: Corticostriatal Determinants of Goal-Directed and Habitual Action. *Neuropsychopharmacology* 35 (1): 48–69. https://doi.org/10.1038/npp.2009.131.

Balleine, B.W., A.S. Killcross, and A. Dickinson. 2003. The Effect of Lesions of the Basolateral Amygdala on Instrumental Conditioning. *Journal of Neuroscience* 23 (2): 666–675.

Baron, J., and I. Ritov. 2009. The Role of Probability of Detection in Judgments of Punishment. *SSRN Electronic Journal* 1 (2): 553–590. https://doi.org/10.2139/ssrn.1463415.

Berker, S. 2009. The Normative Insignificance of Neuroscience. *Philosophy & Public Affairs* 37 (4): 293–329. https://doi.org/10.1111/j.1088-4963.2009.01164.x.

Berkowitz, L. 1989. Frustration-Aggression Hypothesis: Examination and Reformulation. *Psychological Bulletin* 106 (1): 59–73.

———. 2012. A Different View of Anger: The Cognitive-Neoassociation Conception of the Relation of Anger to Aggression. *Aggressive Behavior* 38 (4): 322–333. https://doi.org/10.1002/ab.21432.

Berman, M. 2011. Two Types of Retributivism. In *The Philosophical Foundations of Criminal Law*, ed. S. Duff and R.A. Green. Oxford: Oxford University Press.

Bermúdez, J.L. 2003. *Thinking Without Words*. New York: Oxford University Press.

Berridge, K.C. 1996. Food Reward: Brain Substrates of Wanting and Liking. *Neuroscience and Biobehavioral Reviews* 20 (1): 1–25.

Blanchard, D.C., and R.J. Blanchard. 1984. Affect and Aggression: An Animal Model Applied to Human Behavior. In *Advances in the Study of Aggression*, ed. R.J. Blanchard and D.C. Blanchard, vol. 1, 1–62.

Buckholtz, J., J. Martin, M. Treadway, and K. Jan. 2015. From Blame to Punishment: Disrupting Prefrontal Cortex Activity Reveals Norm Enforcement Mechanisms. *Neuron* 87 (6): 1369–1380.

Carlsmith, K.M. 2006. The Roles of Retribution and Utility in Determining Punishment. *Journal of Experimental Social Psychology* 42 (4): 437–451. https://doi.org/10.1016/j.jesp.2005.06.007.
———. 2008. On Justifying Punishment: The Discrepancy Between Words and Actions. *Social Justice Research* 21 (2): 119–137. https://doi.org/10.1007/s11211-008-0068-x.
Carlsmith, K.M., and J.M. Darley. 2008. Psychological Aspects of Retributive Justice. *Advances in Experimental Social Psychology* 40 (07): 193–236. https://doi.org/10.1016/S0065-2601(07)00004-4.
Carlsmith, K.M., J.M. Darley, and P.H. Robinson. 2002. Why Do We Punish?: Deterrence and Just Deserts as Motives for Punishment. *Journal of Personality and Social Psychology* 83 (2): 284–299. https://doi.org/10.1037//0022-3514.83.2.284.
Clutton-Brock, T.H., and G.A. Parker. 1995. Punishment in Animal Societies. *Nature* 373 (19): 209–216.
Crockett, M.J. 2013. Models of Morality. *Trends in Cognitive Sciences* 17 (8): 363–366. https://doi.org/10.1016/j.tics.2013.06.005.
Cushman, F. 2013. Action, Outcome, and Value: A Dual-System Framework for Morality. *Personality and Social Psychology Review* 17 (3): 273–292. https://doi.org/10.1177/1088868313495594.
Daly, M., and M. Wilson. 1988. *Homicide*. New Brunswick: Transaction Publishers.
Darley, J.M., K.M. Carlsmith, and P.H. Robinson. 2000. Incapacitation and Just Deserts as Motives for Punishment. *Law and Human Behavior* 24 (6): 659–683.
Day, J., R. Wheeler, and M. Roitman. 2006. Nucleus Accumbens Neurons Encode Pavlovian Approach Behaviors: Evidence from an Autoshaping Paradigm. *European Journal of Neuroscience* 23 (5): 1341–1351.
Dayan, P., and K.C. Berridge. 2014. Model-Based and Model-Free Pavlovian Reward Learning: Revaluation, Revision, and Revelation. *Cognitive, Affective, & Behavioral Neuroscience* 14 (2): 473–492. https://doi.org/10.3758/s13415-014-0277-8.
Dayan, P., and N.D. Daw. 2008. Decision Theory, Reinforcement Learning, and the Brain. *Cognitive, Affective, & Behavioral Neuroscience* 8 (4): 429–453. https://doi.org/10.3758/CABN.8.4.429.
de Quervain, D.J.-F., U. Fischbacher, V. Treyer, M. Schellhammer, U. Schnyder, A. Buck, and E. Fehr. 2004. The Neural Basis of Altruistic Punishment. *Science (New York, N.Y.)* 305 (5688): 1254–1258. https://doi.org/10.1126/science.1100735.
Dickinson, A., and B. Balleine. 1994. Motivational Control of Goal-Directed Action. *Animal Learning & Behavior* 22 (1): 1–18.
Ekman, P., and D. Cordaro. 2011. What is Meant by Calling Emotions Basic. *Emotion Review* 3 (4): 364–370. https://doi.org/10.1177/1754073911410740.
Fehr, E., and S. Gächter. 2002. Altruistic Punishment in Humans. *Nature* 415 (6868): 137–140. https://doi.org/10.1038/415137a.
Frank, R.H. 1988. *Passions Within Reason: The Strategic Role of the Emotions*. New York: Norton. https://doi.org/10.2307/2072516.
Gintis, H., J. Henrich, S. Bowles, R. Boyd, and E. Fehr. 2008. Strong Reciprocity and the Roots of Human Morality. *Social Justice Research* 21 (2): 241–253. https://doi.org/10.1007/s11211-008-0067-y.
Gläscher, J., Hampton, A. N., & O'Doherty, J. P. 2009. Determining a role for ventromedial prefrontal cortex in encoding action-based value signals during reward-related decision making. *Cerebral Cortex* 19 (2): 483–495. https://doi.org/10.1093/cercor/bhn098.
Gläscher, J., N. Daw, P. Dayan, and J.P. O'Doherty. 2010. States Versus Rewards: Dissociable Neural Prediction Error Signals Underlying Model-Based and Model-Free Reinforcement Learning. *Neuron* 66 (4): 585–595. https://doi.org/10.1016/j.neuron.2010.04.016.
Goldberg, J.H., J.S. Lerner, and P.E. Tetlock. 1999. Rage and Reason: The Psychology of the Intuitive Prosecutor. *European Journal of Social Psychology* 29: 781–795.

Greene, J.D. 2008. The Secret Joke of Kant's Soul. In *Moral Psychology, Vol. 3, The Neuroscience of Morality: Emotion, Disease, and Development*, ed. W. Sinnott-Armstrong, 35–80. Cambridge: MIT Press.

———. 2009. The Cognitive Neuroscience of Moral Judgment. *The Cognitive Neurosciences* 4: 987–999.

———. 2015. Beyond Point-and-Shoot Morality: Why Cognitive (Neuro) Science Matters for Ethics. *The Law & Ethics of Human Rights* 9 (2): 141–172.

Greene, J.D., and J. Haidt. 2002. How (and Where) Does Moral Judgment Work? *Trends in Cognitive Sciences* 6 (12): 517–523.

Greene, J.D., R.B. Sommerville, L.E. Nystrom, J.M. Darley, and J.D. Cohen. 2001. An fMRI Investigation of Emotional Engagement in Moral Judgment. *Science (New York, N.Y.)* 293: 2105–2108. https://doi.org/10.1126/science.1062872.

Greene, J.D., L.E. Nystrom, A.D. Engell, J.M. Darley, and J.D. Cohen. 2004. The Neural Bases of Cognitive Conflict and Control in Moral Judgment. *Neuron* 44: 389–400.

Haidt, J. 2001. The Emotional Dog and Its Rational Tail: A Social-Intuitionist Approach to Moral Judgment. *Psychological Review* 108: 814–834.

Hamlin, J.K. 2014. Context-Dependent Social Evaluation in 4.5-Month-Old Human Infants: The Role of Domain-General Versus Domain-Specific Processes in the Development of Social Evaluation. *Frontiers in Psychology* 5: 614. https://doi.org/10.3389/fpsyg.2014.00614.

Hamlin, J.K., K. Wynn, P. Bloom, and N. Mahajan. 2011. How Infants and Toddlers React to Antisocial Others. *Proceedings of the National Academy of Sciences of the United States of America* 108 (50): 19931–19936. https://doi.org/10.1073/pnas.

Henrich, J., R. McElreath, A. Barr, J. Ensminger, C. Barrett, A. Bolyanatz, et al. 2006. Costly Punishment across Human Societies. *Science* 312 (5781): 1767–1770. https://doi.org/10.1126/science.1127333.

Kahane, G. 2012. On the Wrong Track: Process and Content in Moral Psychology. *Mind & Language* 27 (5): 519–545. https://doi.org/10.1111/mila.12001.

Kumar, V., and R. Campbell. 2012. On the Normative Significance of Experimental Moral Psychology. *Philosophical Psychology* 25 (3): 311–330. https://doi.org/10.1080/09515089.2012.660140.

Lerner, J.S., J.H. Goldberg, and P.E. Tetlock. 1998. Sober Second Thought: The Effects of Accountability, Anger, and Authoritarianism on Attributions of Responsibility. *Personality and Social Psychology Bulletin* 24 (6): 563–574.

Li, J., M.R. Delgado, and E.A. Phelps. 2011. How Instructed Knowledge Modulates the Neural Systems of Reward Learning. *Proceedings of the National Academy of Sciences of the United States of America* 108 (1): 55–60. https://doi.org/10.1073/pnas.1014938108.

McCullough, M.E., R. Kurzban, and B.A. Tabak. 2012. Cognitive Systems for Revenge and Forgiveness. *The Behavioral and Brain Sciences* 36 (1): 1–15. https://doi.org/10.1017/S0140525X11002160.

McDowell, J. 1982. Reason and Action—III. *Philosophical Investigations* 5 (4): 301–305. https://doi.org/10.1111/j.1467-9205.1982.tb00502.x.

Moore, M.S. 2010. *Placing Blame: A Theory of the Criminal Law*. Oxford: Oxford University Press.

Nadelhoffer, T., S. Heshmati, D. Kaplan, and S. Nichols. 2013. Folk Retributivism and the Communication Confound. *Economics and Philosophy* 29 (02): 235–261. https://doi.org/10.1017/S0266267113000217.

Nelissen, R.M.A., and M. Zeelenberg. 2009. Moral Emotions as Determinants of Third-Party Punishment: Anger, Guilt, and the Functions of Altruistic Sanctions. *Judgment and Decision making* 4 (7): 543–553.

Nisbett, R., and T. Wilson. 1977. Telling More than We Can Know: Verbal Reports on Mental Processes. *Psychological Review* 84 (3): 231–259.

Pollock, J.L. 1987. Defeasible Reasoning. *Cognitive Science* 11 (4): 481–518. https://doi.org/10.1207/s15516709cog1104_4.

Quinn, W. 1985. The Right to Threaten and the Right to Punish. *Philosophy & Public Affairs* 14 (4): 327–373.

Rawls, J. 1955. Two Concepts of Rules. *The Philosophical Review* 64 (1): 3–32. https://doi.org/10.2307/2182230.

Richerson, P., and R. Boyd. 2005. *Not by Genes Alone: How Culture Transformed Human Evolution*. Chicago: University of Chicago Press.

Schultz, W., and P. Apicella. 1992. Neuronal Activity in Monkey Ventral Striatum Related to the Expectation of Reward. *The Journal of Neuroscience* 12 (12): 4595–4610.

Seymour, B., T. Singer, and R. Dolan. 2007. The Neurobiology of Punishment. *Nature Reviews. Neuroscience* 8 (4): 300–311. https://doi.org/10.1038/nrn2119.

Sheffield, F.D. 1965. Relation Between Classical Conditioning and Instrumental Learning. In *Classical Conditioning*, ed. W.F. Prokasy, 302–322. New York: Appleton-Century-Crofts.

Sher, G. 1987. *Desert*. Princeton: Princeton University Press.

Sinnott-Armstrong, W. 2003. Consequentialism. In *Stanford Encyclopedia of Philosophy*. Center for the Study of Language and Information, Stanford University. https://plato.stanford.edu/entries/consequentialism/

Smith, M. 1987. The Humean Theory of Motivation. *Mind* 96 (381): 36–61. https://doi.org/10.1093/mind/XCVI.381.36.

Strobel, A., J. Zimmermann, A. Schmitz, M. Reuter, S. Lis, S. Windmann, and P. Kirsch. 2011. Beyond Revenge: Neural and Genetic Bases of Altruistic Punishment. *NeuroImage* 54 (1): 671–680. https://doi.org/10.1016/j.neuroimage.2010.07.051.

Tindell, A.J., K.S. Smith, K.C. Berridge, and J.W. Aldridge. 2009. Dynamic Computation of Incentive Salience: "Wanting" What Was Never "Liked". *The Journal of Neuroscience* 29 (39): 12220–12228. https://doi.org/10.1523/JNEUROSCI.2499-09.2009.

Wiegman, I. 2017. The Evolution of Retribution: Intuitions Undermined. *Pacific Philosophical Quarterly* 98 (2): 193–218. https://doi.org/10.1111/papq.12083.

Wiegman, I. 2019. Payback without bookkeeping: The origins of revenge and retaliation. *Philosophical Psychology* 32 (7): 1100–1128. https://doi.org/10.1080/09515089.2019.1646896.

Yin, H.H., B.J. Knowlton, and B.W. Balleine. 2004. Lesions of Dorsolateral Striatum Preserve Outcome Expectancy But Disrupt Habit Formation in Instrumental Learning. *European Journal of Neuroscience* 19 (1): 181–189. https://doi.org/10.1111/j.1460-9568.2004.03095.x.

Chapter 7
Normative Implications of Neuroscience and Sociobiology – Intended and Perceived

Ullica Segerstrale

Abstract This chapter discusses the potential social consequences of scientific claims about human behavior. This has been an important concern in major academic "nature-nurture" debates, including the sociobiology controversy around Harvard biologist E. O. Wilson. Neuroscience is the field that Wilson hoped would continue his work in *Sociobiology* (Wilson, Sociobiology: The new synthesis, Harvard University Press, Cambridge, MA, 1975) in regard to finding an evolutionary basis for morality, and today's neuroscience is well positioned to try to fulfill his vision. Its results seem consistent e.g., with Daniel Kahneman's (Thinking fast and slow. Farrar, Strauss & Giroux, New York, 2011) famous "fast and slow" thinking, and social psychology. However, neuroscience is vulnerable to the same criticism as sociobiology: the danger of normative interpretations of factually intended statements. Humans do tend to reason from facts to values, and tentative results often come to serve policy purposes (Segerstrale, Defenders of the truth: the battle for science in the sociobiology debate and beyond. Oxford University Press, Oxford, 2000). In the present social situation, I am particularly worried about the use of the term "tribal" and the consequences of scientists presenting ingroup-outgroup conflict as normal and necessary (Greene, Joshua, Moral tribes: emotion, reason, and the gap between us and them. Penguin Press, New York, 2013; Wilson, The social conquest of earth. Liveright Publishing Company (W. W. Norton), New York, 2012a; Wilson, Evolution and our inner conflict. The opinion pages, The New York Times, June 24, 2012b).

Keywords fMRI · Ingroup-outgroup · Us and them · Tribal · Morality · Facts and values · Trolley problem research · Unintended consequences

---

U. Segerstrale (✉)
Department of Social Sciences, Illinois Institute of Technology, Chicago, IL, USA
e-mail: segerstrale@iit.edu

G. S. Holtzman, E. Hildt (eds.), *Does Neuroscience Have Normative Implications?*, The International Library of Ethics, Law and Technology 22,
https://doi.org/10.1007/978-3-030-56134-5_7

7.1 Introduction

Neuroscience is one of those scientific fields which is hoped to yield answers to fundamental human questions. It was in neuroscience that E. O. Wilson envisioned that the discipline of sociobiology would find its ultimate explanation, in a final "consilience" of the social and natural sciences. He might be classified as a believer in naturalized ethics – the idea that moral principles can be derived from scientific facts. Wilson's most radical formulations appear in his paper with philosopher Michael Ruse (Ruse and Wilson 1986), "Moral Philosophy as Applied Science", which declares "No abstract moral principles exist outside the particular nature of individual species" (p. 186), and talks about the need "to escape – not a minute too soon – from the debilitating absolute distinction between *is* and *ought*" (p. 174). So from the point of view of naturalized ethics, neuroscience would indeed have direct normative implications. Among moral philosophers relatively few support the idea of a naturalized ethics, although there is great interest in the relatively new technique of functional magnetic resonance imaging (fMRI).

Joshua Greene, the current director of the moral cognition lab at Harvard uses fMRI to illuminate important problems in moral philosophy and moral psychology. He answers the question whether neuroscience has normative implications as follows:

> "Whereas I am skeptical of attempts to derive moral principles from scientific facts, I agree with the proponents of naturalized ethics that scientific facts can have profound moral implications, and that moral philosophers have paid too little attention to relevant work in the natural sciences….My understanding of the relationship between science and normative ethics, however, is different from that of naturalized ethicists….Their aim is to find theories of right and wrong that in some sense match natural human practice. By contrast, I view science as offering a 'behind the scenes' look at human morality….*the scientific investigation of human morality can help us to understand human moral nature, and in so doing change our opinion of it*" (Greene 2003, italics added).

This, then, is what he set out to do in his research. Using fMRI he investigated the brain regions correlated with the responses of research subjects as they were being asked to make moral decisions in a set of imagined situations. Here he used the repertory of the existing field of trolley problem research, asking subjects what they thought was the right action to take in regard to variants of a famous (deliberately extreme) model case of a run-away trolley. In the classical version the trolley would be sure to kill five people further down the track unless stopped by either pulling a switch to deviate the trolley or by stopping the trolley by pushing a fat man standing on a footbridge onto the tracks. But pulling the switch would result in killing one person on another track, while pushing the fat man would stop the trolley and thus save five persons.

It was realized that this dilemma case (and its variants) could now be used for neuroscientific exploration – for instance, it might reveal a systematic connection of particular answers with the activation of brain regions with known associations to either emotion or deliberation. Greene's studies led him to conclude that there was indeed a pattern – most people decided to save the five by pulling the switch and sacrifice the single individual, but refused to do so by pushing the fat man (Greene 2013, especially Chap. 5).

Greene was hoping to use the trolley problem to find answers to the long-standing question of the prevalence of either deontology or utilitarianism in our moral reactions. But as it turned out, he found evidence for *both* types of moral reasoning. Inspired by David Kahneman's famous findings about the existence of two modes of thinking, fast and slow (summarized in Kahneman 2011) Greene developed a theory comparing our moral minds to a camera with two settings, a fast automatic one connected to intuitive emotional reactions, and a slower one related to deliberative reasoning. He connected these two settings, respectively, to a deontological and a utilitarian moral processing style. Moreover, he realized that the different styles are triggered by different types of dilemmas, 'personal' or 'impersonal' ones. Pushing the fat man felt personal and therefore brought forth an automatic emotion-driven response.

In his book, *Moral Tribes, Emotion, Reason, and the Gap between Us and Them* (Greene 2013) Greene sets out to examine how the tension between emotion and reason, or deontology and utilitarianism, plays out in larger social settings.

In general Greene thinks that we should not just trust our emotions, we need to involve the slow mode of rational deliberation as well. Our intuitive emotional responses were adaptive in an earlier small-scale evolutionary context, but our situation is different today. Greene himself is on the side of a utilitarian type decision making process, especially when it comes to more abstract policy decisions – there our gut instincts do not provide proper guidance.

So what kinds of reactions did we evolve in regard to social life in bigger groups - and are these appropriate for the complex world of today? He especially identifies a phenomenon that he calls the Tragedy of Commonsense Morality. This refers to our tendency to support the views and interests of "our" group against those of other groups, which results in irreconcilable situations and groups competing with one another. At the same time, he points out, the very cooperation within our own group is dependent on competition with an outgroup.

The Tragedy of the Commons (Hardin 1968) was resolved by people realizing that common resources need to be rationed and selfishness curbed in favor of the long-term interest of the group, but there is not yet any such overriding realization in regard to competing groups which are each convinced that their own position is the right one. We need a moral metatheory to resolve this problem. Greene hopes to convince us about a better way to make moral decisions based on some of his research. He warns us that we should not trust our moral intuitions to give the correct moral guidance but instead aim for "taking it slow" and use deliberative reasoning. This is his well-meaning explicit message to his readers. But I am concerned about a possible *implicit* message that he may be unwittingly conveying in his book as he invites his readers to follow him on his various explorations. What will the readers make of some of the theories and scenarios that Greene brings up *on the way* to his eventual promised metatheory? How will those be received? Will his readers interpret them as carrying a message about how things "naturally" are in the real world – and if so, will this give them normative weight, as well? This is a sincere question on my part, since my long-standing research into nature-nurture controversies, especially the fierce one about sociobiology, shows how easily people treat descriptively intended statements as if they were normative (Segerstrale 2000). That is therefore the meaning that I am giving to the question "Does neuroscience have normative implications?"

7.2 How Facts May Become Values – The Case of the Sociobiology Controversy

Let's take a look at the sociobiology controversy, raging from 1975 to about 2000. The main culprit here was Harvard evolutionist E. O. Wilson's book *Sociobiology: The New Synthesis,* and particularly its last chapter on humans. A group of academic critics, several from Harvard (the Sociobiology Study Group, which later joined the organization Science for the People), claimed that Wilson's aim with his last chapter was political: he wanted to legitimize social inequality by advocating a biological determinist view of human nature (if everything is biologically determined there is no point with social reforms). The group of critics was so sure that *Sociobiology's* message was political that one of the leaders wrote a letter to *Science* telling its readers to "see for themselves", adding "there is politics aplenty in *Sociobiology*, and we who are its critics did not put it there" (Alper et al. 1976). (Incidentally, here the critics relied solely on textual analysis - not the actual politics of leading sociobiologists). Wilson himself actually wanted to convey to a broader public what he saw as exciting new scientific information about social behavior, including the biological underpinnings of many human traits, like morality, believed to be purely cultural. The critics saw this as anathema.

This was not the first time something like this happened. Over the last half century or so fields such as IQ research and behavioral genetics have also come under fire. Critics have feared that biological statements about human behavior will inevitably be exploited for discriminatory social policy or used as justifications for bad individual behavior ("my genes made me do it"). At the same time, the criticized researchers have seen themselves as doing regular science in their fields (and indeed published in respected professional journals).

One way of describing these episodes would be to say that they dealt with the (perceived) consequences of the (perceived) political implications of (purported) scientific facts about the biological foundations of human behavior. It was all in the eye of the beholder - and the social climate was such that the beholder's view was able to prevail for quite a time.

Here is a representative statement by one of the leading critics of sociobiology and IQ research, Wilson's Harvard biology colleague Richard Lewontin as he was interviewed in a local Harvard University newspaper at the beginning of the sociobiology controversy (Lewontin 1975):

> At present our ignorance on this question is so enormous, our investigatory techniques so primitive and weak, our theoretical concepts so unformed, that it is unimaginable to me that lasting, serious truths about human nature are possible. On the other hand the need of the socially powerful to exonerate their institutions of responsibility for the problems they have created is extremely strong. Under these circumstances any investigations into the genetic control of human behaviors is bound to produce a pseudo-science that will inevitably be misused.

Practically nobody dared to defend the attacked scientists in the IQ and sociobiology controversies in public. At the time the "official" belief was that humans were largely cultural (almost blank slates), and also that humans were quite different from animals because of culture. It was taboo to say anything else – for a very long time. The taboo was effectively broken only around the time of the sequencing of the human genome (see Segerstrale 2000).

What was it that Wilson himself wanted with *Sociobiology*? Wilson had a highly unusual agenda: a grandiose long-term plan for science and mankind. Epistemologically, his big ambition was to unite the social and natural sciences around sociobiology. Underlying this was a noble goal: to secure the future of mankind and life on Earth. With the help of biological insights into the truth of human nature, we would be able to make wiser choices and steer away from unfeasible cultural courses, perhaps even self-destruction. But because of the critics' strong counter-interpretation of his words Wilson could not initiate a discussion about these issues.

Not surprisingly, the sociobiologists from the very beginning pointed out that they had been misconstrued - Darwinism is not "advocating" anything, and values cannot be derived from nature! Of course, this may be true - from a strictly logical point of view. We all know about the naturalistic fallacy. But by what mechanism, then, does a factual scientific statement become a seeming prescription for action? The answer is: in a society where people perceive an intimate connection between a fact and its utility. Under such conditions a statement of fact is never really a "mere" statement of fact. And here we have an explanation of the constant concern of the critics of sociobiology – the potential abuse of biological claims. Just as Wilson was interested in positive utility of his field, they were interested in its negative utility. Pointing to historical precedents: racist skull measurements, early IQ research, eugenics, etc., all supposedly based on the newest science (see e.g., Gould 1981), the critics were not convinced that their society would be able to prevent abuse of new biological explanations of humans.

I have argued here that statements of (presumed) facts in evolutionary biology or any other field will turn into political or normative prescriptions as soon as you believe that scientific facts will (or must) be acted upon! (And this may be a tendency in the reasoning not only of some scientists, but also of members of the general public, looking for guidelines for their lives). From the critics' point of view, therefore, certain things should simply not be said. (Indeed, they saw their own role as "weeders" of "bad science", which would necessarily lead to bad consequences; see Segerstrale 2000, Chap. 11).

This is why Wilson's statement about small existing sex differences between men and women (in his book *On Human Nature,* 1978) raised such hackles. Nobody paid attention to his immediately following caveat that society will of course decide how to handle this information, if true (ignore it, enhance it, or counteract it). For the critics, the important point was that "he said it", I learnt from a leading critic who told me that he himself would never have said it (Segerstrale 2000, p. 199).

An indication that the critics saw a strong connection between facts and normative implications is that they applied this type of reasoning not only to sociobiological claims but to themselves as well! Here is the unexpected response I got from a leading Science for the People member in interview when I asked him what he would do if there were ever incontrovertible facts about racial differences. I had expected him to blankly deny that such facts would ever be found. But instead he said: "Then I would evidently have to become a racist!" (He added that there were no such facts yet). (Segerstrale 2000, p. 223).

7.3 Two Tragedies About Morality – And a Potential Third

At the beginning of *Moral Tribes* Greene tells a story explaining how morality came about as a solution to the problem of ruthless self-interest, that famous Tennyson description of a world "red in tooth and claw" (p. 23). His leading metaphor is the Tragedy of the Commons. A group of selfish herders will try to bring as many animals as possible to graze on their commons, a free resource, which will eventually lead to the depletion of resources and everyone's ruin. Moral herders, in contrast, will take into account the total situation and restrict their animal numbers in the interest of everybody. "Thus a group of moral herders, through their willingness to put Us before Me, can avert the Tragedy of the Commons and prosper." (p. 23). For Greene, therefore: "Morality is a set of psychological adaptations that allow otherwise selfish individuals to reap the benefit of cooperation." (p. 23) But our tendency to cooperate comes with an important caveat. "[H]umans were designed for cooperation, but only with some people. Our moral brains evolved for cooperation *within* groups, and perhaps only within the content of personal relationships. Our moral brains did not evolve for cooperation *between* groups (at least not all groups)." (p. 23, italics added).

This is due to the inherent competitiveness of evolution as a process, which makes universal cooperation inconsistent with evolution by natural selection. Evolution is based on competition (pp. 23–24). At the same time, without competition, there would be no cooperation. "[C]ooperative tendencies cannot evolve (biologically) unless they confer a competitive advantage on the cooperators." (p. 24) (In the herder case, the group of morally minded, cooperative individuals from their successful commons will outcompete the group of less morally minded individuals from their failing commons, p. 24). Greene summarizes the situation as follows:

> "Thus, if morality is a set of adaptations for cooperation, we today are moral beings only because our morally minded ancestors outcompeted their less morally minded neighbors. And thus, insofar as morality is a biological adaptation, *it evolved not only as a device for putting Us ahead of Me, but as a device for putting Us ahead of Them*." (p. 24, italics added)

Greene admits that the claim that morality evolved as a device for intergroup competition may sound strange, since it is not obvious how to connect various aspects of individual morality to intergroup competition (attitudes to various things, say abortion or capital punishment). But the connection can be indirect, and our morality can make us do things that go against the forces that gave rise to it – here he refers to Wittgenstein's "evolutionary ladder," which we first climb and then kick away. (p. 25).

So there are actually two tragedies threatening us humans. The traditional Tragedy of the Commons, which pits the individual against the group, Me vs. Us, needs complementing by what Greene calls the modern Tragedy of Commonsense Morality, which puts Us against Them. In other words, the same moral thinking that promotes cooperation within groups serves to undermine cooperation between groups. (Here Greene introduces 'tribe' instead of 'group'). In *Moral Tribes*, therefore, Greene is in search of a workable "metamorality", a higher level thinking which would be able to resolve problems between groups with conflicting moralities and allow them to live peacefully together – just as cooperation was the solution to the Tragedy of the Commons.

What I am interested in, however, is a possible *third* tragedy – the implicit message that Greene may be conveying to the readers (or even mere skimmers) of his book while he is sincerely looking for an answer to the problem that he has identified. Here we have a researcher stating upfront what he takes to be the scientific truth and what he sees as reasonable assumptions about morality. But how might his readers interpret what he is telling us? I here see a parallel to what happened in the sociobiology controversy during the last quarter of the last century.

7.3.1 Moral Tribes *Meets* The Selfish Gene

Let's make a quick overview. Greene believes that "tribal" beliefs are connected to deep values, which explains why different "tribes" do not see eye to eye. In other words, although the Tragedy of the Commons has been overcome by people realizing that everyone is better off by scaling down one's short-term self interest in the interest of the long-term common good, the Tragedy of Commonsense Morality is a harder nut to crack. It is not so clear what could or should be appealed to in this case, since each tribe is operating from "inside" its own moral system, which it sees as naturally correct.

Not surprisingly, the answer lies at a higher level. Greene's solution is to propose the development of a "meta-morality" which would be able to arbitrate between the different moral systems, while taking everybody's interest and concern into account. It would involve a kind of utilitarian calculus, which would help generate possible solutions that everyone would be able to agree on. This system would be based on reason, rather than emotion, and the role of reason would be to strike the right balance between emotion-driven individual interest and reason-driven interest in the good of the group.

Note that in general, Greene believes that our utilitarian calculus, which is reason driven, is the robust type of moral foundation, while differences in individual moral convictions, which are emotionally grounded, are best seen as biased or distorted. The latter are comparable to the famous Muller-Lyer illusion, Greene argues. And because the Muller-Lyer test is known to be an illusion, Greene feels justified to argue for reason and utilitarian calculus as being primary. Compare this to the alternative view that our emotional reactions are primary and any moral rules invoked to explain our behavior are mere rationalizations (see for instance Haidt's famous article on the tail wagging the dog, Haidt 2001).

Moreover, Greene is making the explicit point that the great capability we humans have for cooperation is directly dependent on inter-group competition. Inter-group competition brings out virtuous behavior. In this he joins many others, starting with Darwin himself and ending with E. O. Wilson, who in a number of recent publications has drawn attention to just this ingroup-outgroup opposition. It almost looks as if Wilson wished to launch some kind of simple mantra or slogan as he recently with D.S. Wilson campaigned for restoring the validity of the theory of group or multi-level selection. One article ended with the explicit slogan that selfishness beats altruism within groups, but altruistic groups beat selfish groups (Wilson and Wilson 2007). And what might be the perceived message of a chapter subheading that reads "Tribalism is A Fundamental Human Trait" (Wilson 2012a, p. 57)?

It is these kinds of statements that I now want to take issue with. We saw earlier that Wilson was violently attacked in the sociobiology controversy when he was suggesting things about humans that he as a biologist believed were true - for instance, the existence of small differences between the sexes in regard to preference of profession. But even though he immediately added that biology is not destiny; it is up to society to decide what to do about it – suppress differences, enhance them, or do nothing – this follow-up statement of his was completely ignored and Wilson was labeled a sexist and biological determinist. He was labeled a sexist simply because "he said it" (the statement about sex role differences, always incendiary).

So, in the same way, it seems to me that Joshua Greene is "saying things" in *Moral Tribes*. At the beginning of his book he emphasizes the great depth of the differences between "tribes" of various kinds. The differences go so deep that any reconciliation is unthinkable. That looks like an alarming description. But perhaps Greene believes that he can make such strong statements, because he is working toward solving the bigger problem of the Tragedy of Commonsense Morality and tribal conflict.

But what about the unintended metamessages that the book might be sending on the way toward his problem solution (his meta-morality)? What if Greene's readers, or even those who only know the catchy title of his book, *Moral Tribes*, perceive those bold early statements as a true description of the way things are, or even the way that things are meant to be? They will learn that moral differences are so deep that tribal conflict is inevitable, but the good thing is that such a situation promotes cooperation. In fact, conflict with outgroups is necessary for ingroup cooperation. In this way, what was intended as an early description of a state of affairs, for which Greene planned a solution later on in the book (his meta-morality) might easily take on a normative charge, for people who get impressed by those early strong descriptions (or just the book title) rather than faithfully following Greene through his book on his painstaking exploration and complex justification of his desired metatheoretical solution.

I imagine the situation as somewhat similar to people's reaction to *The Selfish Gene* by Richard Dawkins (1976). For Dawkins, his title captured the new idea of focusing on the gene instead of the organism in evolutionary theorizing. "Selfishness" was here anthropomorphically related to a gene's "interest" in propagating itself, having nothing to do with human selfishness, it was intended as a pedagogical device. Some academics, however, reacted very strongly negatively to this title, not only the group of critics of sociobiology, but also someone like Karl Popper (see Segerstrale 2000, p. 74). Others took it as face value as a catchy title, but there was justified worry that "the innocent layman" would miss the point and think it described exactly the way we are. Much later, Mark Ridley, Dawkins' student and a popular author in his own right, made the obvious countermove by calling his book *The Cooperative Gene* (Ridley 2001).

7.4 Where Have All the Critics Gone?

Considering the storm around sociobiology, it is interesting to note that there appears to have been almost no moral/political protest against neuroscience. There is no organized opposition – no "Neurobiology Study Group", no demonstrations, and no political accusations. This is surprising, since, if anything, neurobiology represents a culmination of the sociobiological project. For Wilson, neuroscience was the direction in which sociobiology was going – the mind was the final frontier.

Is there, then, something significantly different about neuroscience when it comes to triggering the critical imagination about dangerous social implications? Is the difference perhaps that we are allowed to see fMRI results, we are shown how various (impressively named) areas of the brain turn different colors and what this means - we are invited "in" in regard to brain research? fMRI results are charismatic, and get immediate interpretations, unlike those invisible hypothetical genes "for" behavior invoked by sociobiologists.

One answer could be that the socio-political climate has changed. People have learnt "gene talk", especially in conjunction with the human genome project (see e.g., Nelkin and Lindee 1995), and new generations may no longer be averse to biological explanations of behavior. What strongly united the earlier critics of sociobiology was the fear of genetic determinism, which they believed would support discrimination and discourage needed social reform.

Now assuming that the general situation is still the same, in the sense that people have a tendency to treat scientists' descriptive statements of human behavior as normative - what could be done to discourage people from believing that in-group-out-group conflict is unavoidable, and as it would seem, actually desirable, since it promotes in-group cooperation? I see this as a pronouncement that legitimizes conflict and war, and I am surprised that so many today seem to buy into it. I think that acceptance of this thesis as an unproblematic truth is dangerous, as is the references to violent scenarios from history and pre-history that often accompanies this thesis. And going back to *Moral Tribes*, my point is that despite the fact that Greene is on the honorable lookout for a solution to the Tragedy of Commonsense Morality (which he locates in a meta-morality), his unproblematic characterization of the Commonsense Morality as deeply tribal may be sending a wrong signal to his readers. As a result, the take home message for readers may not be what he intended – the need for a meta-morality - but rather a reaffirmation and reinforcement of people's existing biases, potentially giving rise to more, rather than less, "tribalism" and social divisiveness in this world.

7.5 The Critic's Potential Tool Kit

This is why I now want to play critic. With an eye to earlier controversies and current developments in the scientific community I can find at least the following ways to criticize what I see as the overemphasis on ingroup-outgroup conflict, of which Greene is not the only scientist culpable today. Let's see what options we have.

1. We could act like the critics of sociobiology, simply declaring that a scientist should not make certain statements at all, if it could even be imagined that these might have negative social consequences.
2. Or we might want to go with Hilary and Steven Rose, tenacious British critics of sociobiology, evolutionary psychology, and brain research, who are concerned with all kinds of negative social and medical consequences of recent scientific developments, including neuroscience, in *Genes, Cells, and Brains* (Rose and Rose 2012).
3. We could criticize Greene's theory or reasoning on philosophical or scientific grounds. This has been attempted by a number of critics and reviewers of his book, and Greene has often thoughtfully responded to their criticism in later articles, sometimes expanding his own reasoning based on their comments (for instance Greene 2014). We might even try, with Berker (2009) to completely undermine Greene's basic thesis, arguing for "the normative insignificance of neuroscience", a critique which Greene regards as flawed (Greene 2014).
4. We could agree with the title of Thomas Nagel's review of *Moral Tribes* that "You Can't Learn About Morality from Brain Scans" (Nagel 2013). His basic point is that assumptions about the weight of emotion vs. reason or Kantianism vs. utilitarianism have necessarily to be made to enable Greene's theory

construction, and the outcome will necessarily be dependent on those assumptions. In Greene's case this means that however his research results are interpreted, and also in regard to the issue of metamorality, the ultimate weight must be given to utilitarianism, Greene's preferred philosophy. Greene sees utilitarianism as sitting on stable ground, while he regards any deontologic views as emotionally informed and therefore tending to be illusory.

5. We might even criticize fMRI studies as such. There is already the question about what the fMRI scans actually measure (they don't measure ongoing neuronal activity, but rather get their information from a signal (BOLD) that indicates changes in the blood oxygenation level) and there are other problems (e.g., Crawford 2008; *Nature Neuroscience*, 2018) (Incidentally, Greene himself says that his thinking is not dependent on fMRI; Greene 2014)
6. Or we could go a step further with *The Neuroskeptic* (2016), which complains that many f-MRI studies are the subject of "p-hacking" (the selection of studies or data to make correlations statistically significant) and that the whole field of brain research is characterized by too much methodological flexibility. Or with *Nature Behavior* (Munafo et al. 2017) which reports on the current crisis in replicability of psychological and brain research.
7. Or we could look more closely into the research by Cikara and Van Bavel (2014), who study ways in which individuals in fact subjectively perceive their ingroups and outgroups. There are many different ways in which group identification can happen. Also, people's conceptions can change.
8. What about taking a close look at the root of the problem: the ingroup-outgroup bias itself? What are the typical *conditions* under which it expresses itself strongly, and what do we know about conditions under which its effect is diminished? What do we know about Us and Them situations in general? What kind of intervention is possible? There are scattered observations in books about social psychology and sociology. The best short overview is probably Chap. 11, "Us and Them", in Robert Sapolsky's *Behave: the Biology of Humans at Our Best and Worst* (2017). There he provides a small summarizing list of measures to "lessen the adverse effects of Us/Them-ing" (p. 422). Here is the list:

   - emphasizing individuation and shared attributes
   - perspective taking
   - more benign dichotomies
   - lessening hierarchical differences, and
   - bringing people together on equal terms with shared goals.

9. Or what about studying alternative theories relating to the ingroup-outgroup problem? This is where I am going next.

7.6 The Normative Implications of Ingroup-Outgroup and Tribe Talk

I wonder at the recent opposition and strife between ingroup and outgroup in the popular writings of various "guru" scientists. It is not clear what aim this serves, except to implicitly strengthen a belief in the inevitable presence - and necessity of - group conflict. This will fortify those readers who are prone to think that what naturally exists is "good" or "right" in their existing belief. And it does matter (especially if you are an important decision maker) if you believe that ingroup-outgroup conflict is inevitable, or you think that the threat from an outgroup is the only way to bring about group cooperation.

Fortunately there are other researchers, especially anthropologists, economists, political scientists, and game theorists, who have been investigating the general conditions under which group cooperation can develop, and has developed, independently of the threat from an outgroup. These studies emphasize our natural human tendencies to work for the good of the group, develop and follow norms, and punish free riders and rule breakers (e.g., Axelrod 1984; Boehm 2012; Henrich and Boyd 2001; Fehr and Gachter 2002).

As mentioned, some scientists have been interested in reviving "group selection" to explain the evolution of cooperation. Group selection in the strict sense involves selection among groups, and requires that the less fit groups "go extinct". Some leading scientists have assumed that this has simply meant killing off the defeated group (e.g., Bowles 2006; Wilson 2012a, b). However, an alternative explanation (with concrete cases) suggests that members of the defeated group may rather get absorbed by the winner and learn their culture by resocialization (Boyd and Richerson 2009). More is being discovered by using genomic data to assess such things as the timing of genetic changes, and human migration patterns.

A new theory of "cultural group selection" suggests that the human propensity for cooperation may in fact have arisen through a process of gene-culture co-evolution, with culture as the driver. Culture can affect genetic evolution by quickly creating a new environment for adaptation and in this way put pressure on the genes – especially in times of rapid environmental or climatic change (Boyd and Richerson 2005, 2009)

Meanwhile the "group extinction" required by group selection theory might also be moved to the realm of culture instead. The variation between groups that is needed to exist for there to be selection (evolution) at all can in fact be purely cultural, having to do with differences in social norms and ways of doing things. Instead of inter-group competition or group extinction the next step in this case can simply be imitation of neighboring groups with "better" cultural ways. There are also cases where after a conflict members of the defeated group get absorbed by the winner and learn their culture by resocialization. Recent examples exist (Boyd and Richerson 2009). Yet another suggestion is that cultural rules may have created a selection pressure for "cooperative" genotypes (Bell et al. 2009).

When it comes to the ingroup-outgroup phenomenon, we know from experimental studies already in the 1970s (e.g., Tajfel 1971) that this opposition is very easily created. What is less well known, or publicized, are the ways to stop an emerging conflict and reconcile opposing parties. Musafer Sherif (1966) investigated this in his famous boys' camp experiment: reconciliation was possible to achieve when two hostile groups had to work together to handle a crisis situation. What we need more of is good knowledge of and research into the variety of circumstances under which ingroup-outgroup oppositions can be – and have been - successfully overcome. We need success stories, and to share them with the general public.

And finally, can we please just stop the "tribe talk"? The word "tribe" seems to be popping up everywhere these days in intellectual and popular discourse. The term itself may be flexibly used, but its very use potentially does two bad things: it legitimizes the division of social reality into conflicting ingroups and outgroups, and it becomes a new handy excuse for bad attitude or behavior - "that is just tribal!"

References

Alper, J., et al. 1976. The Implications of Sociobiology. *Science* 192: 424–425.

Axelrod, R. 1984. *The Evolution of Cooperation*. New York: Basic Books.

Bell, A.V., P.J. Richerson, and R. McElreath. 2009. Culture Rather Than Genes Provides Greater Scope for the Evolution of Large-Scale Human Prosociality. *Proceedings. National Academy of Sciences. United States of America* 106: 17671–17674.

Berker, S. 2009. The Normative Insignificance of Neuroscience. *Philosophy and Public Affairs* 37: 293–329.

Boehm, C. 2012. *Moral Origins: The Evolution of Virtue, Altruism, and Shame*. New York: Basic Books.

Bowles, S. 2006. Group Competition, Reproductive Leveling and the Evolution of Human Altruism. *Science* 314: 1569–1572.

Boyd, R., and P.J. Richerson. 2005. *Not by Genes Alone*. Chicago: University of Chicago Press.

———. 2009. Culture and the Evolution of Human Cooperation. *Philosophical Transactions of the Royal Society B* 364: 3281–3288.

Cikara, M., and J. Van Bavel. 2014. The Neuroscience of Intergroup Relations: An Integrative Review. *Perspectives of Psychological Science* 9 (3): 245–274.

Crawford, M.B. 2008. The Limits of Neuro-Talk. *The New Atlantis* 2008: 65–78.

Dawkins, R. 1976. *The Selfish Gene*. Oxford: Oxford University Press.

Fehr, E., and S. Gachter. 2002. Altruistic Punishment in Humans. *Nature* 415: 137–140.

Gould, S.J. 1981. *The Mismeasure of Man*. New York: W. W. Norton.

Greene, J. 2003. From Neural 'is' to Moral 'Ought': What are the Moral Implications of Neuroscientific Moral Psychology? *Nature Reviews Neuroscience* 4 (10): 846–850.

Greene, Joshua. 2013. *Moral Tribes: Emotion, Reason, and the Gap between Us and Them*. New York: Penguin Press.

Greene, J. 2014. Beyond Point-and-Shoot Morality: Why Cognitive Neuroscience Matters for Ethics. *Ethics* 124 (4): 695–726.

Haidt, J. 2001. The Emotional Dog and Its Rational Tail: A Social Intuitionist Approach to Moral Judgment. *Psychological Review* 108: 814–834.

Hardin, G. 1968. The Tragedy of the Commons. *Science* 162: 143–148.

Henrich, J., and R. Boyd. 2001. Why People Punish Defectors. *Journal of Theoretical Biology* 208: 79–89.

Kahneman, Daniel. 2011. *Thinking Fast and Slow*. New York: Farrar, Strauss & Giroux.
Lewontin, R.C. 1975. Interview. *Harvard Crimson*, December 3, 1975 (cited in Segerstrale, 2000, p. 203).
Munafo, M., et al. 2017. A Manifesto for Reproducible Science. *Nature Human Behavior* 1: 0021.
Nagel, T. 2013. You Can't Learn About Morality from Brain Scans. Review of J. Greene *Moral Tribes*. *New Republic*, November 1, 2013.
Nelkin, D., and M.S. Lindee. 1995. *The DNA Mystique*. New York: Freeman.
Neuroskeptic. 2017. Two Manifestos for Better Science. January 11, 2017 (online).
Ridley, M. 2001. *The Cooperative Gene. How Mendel's Demon Explains the Evolution of Complex Traits*. New York: The Free Press.
Rose, H., and S. Rose. 2012. *Genes, Cells and Brains: The Promethean Promises of the new Biology*. London: Verso.
Ruse, M., and E.O. Wilson. 1986. Moral Philosophy as Applied Science. *Philosophy* 61: 173–192.
Sapolsky, R. 2017. *Behave: The Biology of Humans at Our Best and Worst*. New York: Penguin.
Segerstrale, U. 2000. *Defenders of the Truth: The Battle for Science in the Sociobiology Debate and Beyond*. Oxford: Oxford University Press.
Sherif, M. 1966. *In Common Predicament: Social Psychology of Intergroup Conflict and Cooperation*. Boston: Houghton Mifflin.
Tajfel, H. 1971. Experiments in Intergroup Discrimination. *Scientific American* 223 (5): 96–102.
Wilson, E.O. 1975. *Sociobiology: The New Synthesis*. Cambridge, MA: Harvard University Press.
———. 1978. *On Human Nature*. Cambridge, MA: Harvard University Press.
———. 2012a. *The Social Conquest of Earth*. New York: Liveright Publishing Company (W. W. Norton).
———. 2012b. Evolution and Our Inner Conflict. The Opinion Pages, *The New York Times*, June 24.
Wilson, D.S., and E.O. Wilson. 2007. Rethinking the Theoretical Foundations of Sociobiology. *Quarterly Review of Biology* 82 (4): 327–348.

Chapter 8
Nervous Norms

Matthew Ruble

Abstract Should we think that 'more facts' entails 'better morality'? We do think this way for a great number of contemporary moral issues. After all, forming a moral view in an evidence free manner seems both morally and epistemically vicious. But failing to muster an empirically informed moral view is only one way to err regarding the relationship between empirical facts and moral norms. We might also make the mistake of *overly relying* on facts when engaging moral queries. Such is the mistake that medical ethics, psychiatric ethics, and more recently, neuroethics, commit when attempting to adopt a 'facts first then values' approach employed in medical ethics. This methodology assumes that we begin moral inquiry well equipped with uncontroversial factual evidence only after which we engage contested moral values. Any methodology of arguing from allegedly undisputed facts to disputed values is a methodology doomed to moral and epistemic failure. This mistake has been well documented by KWM Fulford and this chapter attempts to apply this fundamental mistake of psychiatric ethics as a moral and methodological lesson for future neuroethics. This chapter closes with a plea for interdisciplinary teams to be the fundamental unit of research into the normative implications of neuroscience rather than isolated scholars in isolated pre-defined disciplinary specialties.

Keywords Neuroscience · Neuroethics · Psychiatric ethics · Medical ethics · Psychopaths · Function

8.1 Introduction

Two errors, one conceptual and one methodological, both working in concert, pose a challenge that any future neuroethics must come to terms with. We can look into the ways psychiatric ethics has committed this tandem-error in order to help

M. Ruble (✉)
Department: Philosophy and Religion, Apalachian State University, Boone, NC, USA
e-mail: rublemw@appstate.edu

G. S. Holtzman, E. Hildt (eds.), *Does Neuroscience Have Normative Implications?*, The International Library of Ethics, Law and Technology 22,
https://doi.org/10.1007/978-3-030-56134-5_8

neuroethics avoid it.[1] Drawing primarily from the contributions of philosopher and psychiatrist Bill Fulford, this chapter will trace the conceptual debate over 'health,' its counterpart 'disease,' narrowed to a focus on 'function', how mental illness has forced us to rethink these concepts, and what implications this has for the ways we do medical ethics, psychiatric ethics, and neuroethics. In short, the first payoff of this chapter serves as a cautionary tale for neuroethics as a branch of bioethics. The second payoff relates more centrally to the questions surrounding the implications neuroscience may or may not have for our understanding of moral norms by showing where that literature (for example, Greene 2008; Cushman and Young 2009; Koenigs et al. 2007) is reflected in the debate over whether or not psychopaths (individuals diagnosed with Antisocial Personality Disorder) are morally responsible (for example, Watson 1993, 1996; Levy 2007a, b; Nelkin 2015; Talbert 2008, 2011). Both discussions converge on the question of the functioning of the ventromedial prefrontal cortex (vmPFC), one in search of the neural correlates of the optimal normative moral theory, and the other in search of the neural basis for moral responsibility. The core of the error, as I argue, lies in an incorrect assumption that 'function' can be accounted for as a non-normative, empirical/descriptive concept, and this leads to a secondary methodological error that a 'facts first, then values' approach is appropriate for neuroethics. The chapter ends with a practical plea for a neurohumanities to serve as ballast to neuroscience.

8.2 Medical Ethics, Psychiatric Ethics and the Debt to Fulford

The literature surrounding conceptions of 'health,' and the corresponding literature on 'mental health' is vast. This is not the place to recall and summarize the rich and storied conceptual debate over 'health.'[2] Rather, here I wish to focus on the contributions to that debate over mental illness of KWM Fulford's linguistic-analytic approach to medical ethics for two key insights.[3] The first reveals that a conceptual analysis of mental illness is highly evaluative and thus relatively contested compared to its counterpart physical illness. This places values, norms and ethics at the

[1] 'Psychiatric ethics' includes ethical considerations specifically relevant to the classification, diagnosis and treatment of individuals with mental illness (including but not limited to psychiatry, psychology, social services, etc.).

[2] For a very few key articles on health, see Boorse 1975, 1997; Callahan 1973; Daniels 2007; Nordenfelt 1987. For just a few select key articles in the debate over mental illness and psychiatry's contested status as properly belonging to medicine, see Szasz 1963; Kendell 1973; Fulford 1989; Thornton 2000; Radden 2004; Fulford et al. 2013.

[3] Fulford offers two unique advancements to our discussions surrounding medical ethics: the linguistic-analytic approach to philosophy and an emphasis on psychiatric ethics as paradigmatic of, and informative to, medical ethics writ large. See specifically the following works from Fulford 1994, 1998, 2000, 2004.

conceptual core of psychiatry (including classification and diagnosis) and this is quite unlike the empirical sciences serving as the conceptual core of physical medicine.[4] The second contribution of Fulford is his non-descriptivist account of the concept 'function.'[5]

Philosophy of psychiatry has revitalized the conceptual debate over health in part by sharpening focus on 'function' as a descriptive concept that can place psychiatry more securely alongside its empirical medical counterpart. There are promising lessons for neuroethics both as a branch of bioethics and with respect to the value of neuroscience in aiding our understanding of normative moral theory.

The conceptual debate over mental health remains robust and the philosophy of psychiatry has developed along many avenues of conceptual examination.[6] There remains the essential tension between empirical (facts) and normative (values) conceptions of mental health/illness, though now with revised focus, for instance, ranging from broader questions of whether or not mental illnesses are natural kinds (Tekin 2017), or perhaps interactive kinds (Hacking 1995, 1998), to focus on specific, or discrete, mental illnesses (Psychotic Disorders, Depression, Attention Deficit Hyperactivity Disorder, etc.) and how they force us to reconsider core issues in medical ethics. For instance, patients diagnosed with Anorexia Nervosa require us to reconsider the conditions for decision-making capacity to potentially include 'pathological values' when the previous conditions (understanding, weighing-up, and communicating) have been met, and this has direct bearing on the ethical dilemma of respecting patient autonomy or beneficently overriding patient refusal of treatment (Tan et al. 2007). More pertinent to this discussion is the debate over whether or not psychopaths (individuals diagnosed with Antisocial Personality Disorder) suffer from a neurobiological disorder that renders them incapable of moral agency and thus, moral responsibility (Cleckley 1955; Elliot 1996; Charland 2004; Levy 2005, 2007a, b; Nelkin 2015; Jalava and Griffiths 2017). The debate over the moral responsibility of psychopaths appears to hinge considerably on empirical research into the *function* of the vmPFC in psychopaths, and the question of neural function connects directly to the normative implications of neuroscientific research of moral dilemmas (Greene 2008; Cushman and Young 2009; Koenigs et al. 2007). This brings us back to the central concept of 'function' and Fulford's analysis, which will have direct bearing on the approaches to and methodologies of ethics in medicine, psychiatry, and neuroscience.

---

[4] Here I must overlook distinctions between values, norms and ethics for purposes of brevity. Moreover, one might include values, norms and ethics under the title 'humanism' as the (proper) disciplinary core of psychiatry. This of course remains an open question, and is an argument for which I am advancing, and to which I hope to include 'neurohumanities' as a disciplinary extension.

[5] For a rich discussion on 'function' see the special edition of *Philosophy, Psychiatry and Psychology*, 2000, volume 7, number 1. See especially the texts by Thornton, Fulford, Megone, and Wakefield.

[6] Much of this owes to increasing venues and scholarship in the area, most notably the journal *Philosophy, Psychiatry and Psychology* and the renaissance of philosophy of psychiatry owed principally to the work of Bill Fulford.

What is meant by 'methodologies of ethics' in medicine, psychiatry and neuroscience? Very broadly speaking, the methodology of medical ethics, given the conceptual commitment that medicine is at its conceptual core an empirical science, adopts 'facts first, then values' considerations.[7] The question is whether or not psychiatry (and neuroethics) is conceptually positioned to adopt the same 'facts first, then values' ethical methodology. Fulford's argument is that psychiatry is in no such position, and better for psychiatric ethics (and those receiving mental health services) to confront this difficulty head-on. Let's take a closer look at the core of Fulford's argument.

The following is a brief summary of Fulford's (1994, 2000) reflections on the conceptual difficulties in medicine. He describes a pervasive view (one that he criticizes) that although many non-scientific disciplines (sociology, anthropology, law, economics, moral philosophy, etc.) are relevant to medicine, the 'medical model' argues that at its conceptual and disciplinary core medicine is an empirical science for which technical expertise in physiology, anatomy, biochemistry, neuroscience, etc. holds all the future promises. It is not controversial to point out that the current cultural legitimacy of medicine rests on its empirical status (hence the rise of evidence-based medicine), and that this view is widely shared. Fulford rightly notes that this is a naive view of medicine and psychiatry. Nonetheless, the neurobiological model of psychiatry remains entrenched. What is controversial about Fulford's view is that according to Fulford, the medical model appears to marginalize ethical concerns to (at best) a secondary consideration once all the antecedent empirical facts are in place (say, in the form of diagnosis and consideration of treatment options). But psychiatry, owing to the non-empirical influence (social, ethical and legal 'values' or 'norms') on the very classification and diagnosis of mental illness, faces considerable obstacles to adopting the medical model, including what we might call the 'facts first, then values' approach to medical ethics. To be clear, Fulford's concern implies that the 'facts first, then values' model is conceptually and morally problematic for *both* medicine and psychiatry (psychiatry as a specialization within medicine) because values and normative considerations are overlooked or erroneously reduced to, or subsumed by the facts.[8] But psychiatry is particularly vulnerable to normative influences, and at a fundamentally conceptual level, as its aim is to identify and treat disorders of thought and behavior.

Let us consider the preceding observations by Fulford to form the basis of what we might call the conceptual differences argument. I take the following to be a

[7] There are of course narrower debates within medical ethics about the correct normative approach to addressing moral dilemmas - principles-based, casuistry, virtue theoretical approaches, etc. These normative theoretical tensions, (e.g., should we maximize utility even at the expense of disrespecting autonomy?), are conceptually situated, as I have in mind, one tier lower than the more abstract 'facts first, then values' methodology of ethics. My point here is also that all of these theoretical approaches appear to assume that the relevant medical facts are settled and antecedent to ethical considerations. This may be controversial but I here make the assumption to follow the line of present inquiry.

[8] See Fulford (2004, pp. 205–234) for a compelling call to balance evidence-based medicine with values-based medicine.

faithful albeit brief rephrasing and summary of that argument: Because of the diversity of human experience, mental illnesses are normatively loaded both in content, and in practical effect (or upshot) relative to their counterpart in physical illnesses, which are essentially non-normative in content yet normative in practical effect. If this is correct, the argument that at its conceptual core mental illness is normative, then psychiatry is itself a normative intervention, and is not a non-normative essentially empirical discipline. And if psychiatry is normative, then it is not conceptually or epistemically positioned to first consider all the facts prior to engaging in ethical considerations. Hence, the 'facts first, then values' approach appears not to be a methodology appropriate for psychiatric ethics. It may well not be appropriate for medical ethics either, and if so, then there is a lesson to be learned for medical ethics from psychiatric ethics.

Even with Fulford's conceptual differences argument in place we would be hasty without considering a doubling down of the objection that the neurobiological turn in psychiatry does not accept the charge that mental illness is normative in content. Although there are a growing number of voices skeptical of the neurobiological model of mental illness, the explanation of psychopathology as arising etiologically from neuro-anatomical dysfunction of the brain has undeniably gained widespread acceptance.[9] Much then, hinges on the concept of 'function' and its conceptual status with respect to its normative content. This brings us to Fulford's second contribution to our understanding of (mental) health and illness, only now with a narrowed focus on the concept of 'function' and the debate over its normative status.

Fulford (2000, p.78) identifies five key terms involved in the debate over the naturalization of medicine and psychiatry: 'function,' 'dysfunction,' 'disease,' 'illness,' and 'disorder.' These concepts are collectively conceptually linked and form what he calls a logical 'naturalization cascade.' Just where norms and values creep into these concepts remains up for considerable debate. If the concept 'function' can be naturalized (accounted for in straightforwardly empirical and causal terms), so goes the naturalist argument (Wakefield 1992, 1995, 2000, 2009; Kendall 1975), then it can aid in securing science as the basis of medical theory, including psychiatry. This places 'function' as the foundational concept on which the remaining four concepts (dysfunction, disease, illness and disorder) can at least begin to build up the naturalization cascade. But Fulford is deeply skeptical that 'function' can be naturalized, hence serve as the "heuristic holy grail, on which biology, and in turn the theoretical cores of medicine and psychiatry can, in principle, be built up as mature scientific disciplines," (2000, p.83). He argues that 'function' maintains evaluative content owing to the logical closeness of the intrinsic meanings of all the terms in the cascade. And if norms and values are present at any point in the naturalization cascade, then values, however deeply hidden, are present throughout the conceptual cascade.

---

[9] It seems to argue otherwise (whatever form of counterargument) is tantamount to ignoring the empirical evidence on a level equivalent to the climate change denier or flat-earther. See Elliot (2004, 2010) for the biomedical model of psychiatry as a leading contributing factor to the increasing rates and frequency of psychopathology.

To clearly identify the risk before us, what appears to be a distinct kind of disvalue qua biological/medical/psychiatric disvalue (in this case allegedly captured in the term 'dysfunction') is mistaken for another kind of disvalue that is rather moral/social/prudential. The values involved in the latter are concealed in the values involved in the former. Clinical language (such as 'function') appears to smuggle moral and social normative values. If there are indeed distinct kinds of values at play, in both the clinical and the moral language of 'function,' then the logical closeness and reciprocally informing meanings of the concepts used by both camps appears to pollute the entire naturalization cascade with normative values.

We can see Fulford's concern play out before us, for example, by considering the difference between inquiring into proper brain function qua a neuro-biological evaluation of function, with the manner in which 'function' is operative in diagnosing individuals with a psychiatric disorder in the form of the invariant diagnostic criteria (that cut across and thus, connect the discrete disorders) requiring the impairment of social or occupational functioning. Even the naturalist Wakefield (2000, p. 40) is attune to the concern, arguable motivating his naturalism, that we must "rein in the unfettered value-driven application of disorder to anything we dislike." And this renders Fulford's insight all the more important, for as he argues, if we do not remain carefully focused on values operative even in ostensibly scientific biological terms like 'function' we run the risk, at least in psychiatry, not simply of a false objectivity but of moral abuses committed by an ideology mistaken for science (2000, p. 88).

There are two upshots from Fulford's position on functioning I wish to record. The first is that even a biological account of 'function' contains normative content, and second, that we must first (dis)value a given attitude, thought or behavior before we take up the empirical endeavor of seeking out the neuro-anatomical correlates of the (dis)valued behaviors, thoughts and attitudes. The norms then, on Fulford's account, appear to be in the driver's seat:

> "It may well be shown that particular patterns of brain functioning will one day be shown to underlie particular experiences and behaviors. But these patters of brain functioning will only be the causes of *illness*, and hence diagnostically significant, if the experiences and behaviors themselves are first construed as pathological," (1994, p. 194).

If we take Fulford's insights seriously, then psychiatric ethics appears not to be well positioned to adopt the medical model and with it an ethical methodology that assumes a facts first, then values consideration sequence. For the very facts that are alleged to be first order in psychiatry appear to be normatively imbued. The normative content of 'function' with respect to relevance to psychiatry appears to manifest on two levels. The first and more general sense of 'function' is to be found in the ubiquitous diagnostic criteria (often criterion 'A') requiring 'impairment in social or occupational functioning.' This is blatantly socially normative.[10] The second

[10] As captured and maintained in ongoing editions of the *Diagnostic and Statistical Manual of Mental Disorders*, including *DSM 5*, almost all of the 800 plus discrete diagnostic entries in DSM 5 include an iteration of criterion A: 'impairment in social or occupational functioning.' This diag-

manifestation of normative content is the much more specific sense of body part function, and with specific interest in the neurobiological function (vmPFC function, for example). Now we are in a better position to see the impact the preceding analysis has for both branches of neuroethics (the ethics of neuroscience and the neuroscience of ethics).[11]

8.3 Bewitching Correlates, Converging Literatures and Dubious Normative Upshots

Does the same Fulford inspired worry about psychiatry (that ethics remains central to psychiatry, not merely peripheral) extend to neuroscience? I will attempt to make the case that indeed, ethics is not merely peripheral to neuroscience, but appears to lie at its conceptual core. To show this, we now turn, as an example, to the debate over the moral responsibility of psychopaths (individuals with Antisocial Personality Disorder), and how vmPFC function figures into the discussion, thus connecting psychiatric ethics with neuroethics, and thus connecting both to broader questions regarding the normative implications of neuroscience.

The debate over the moral responsibility of psychopaths illustrates the increasingly neurobiological turn in psychiatry in which the symptoms of a given disorder are allegedly explained by neurobiological dysfunction. Before moving on to a discussion of psychopaths there are three broad points I wish to reinforce here. First, *that* the neurobiological model of mental illness owns the current zeitgeist, and second, that the dominance of this model has unfortunately led to both the insularity (Garnar and Hardcastle 2004; Elliot 2004; Tekin 2017) and decreasing weight of other disciplines that are deeply relevant to the understanding and treatment of mental illness. The third point is that philosophers unknowingly reinforce this disciplinary hierarchy by either deliberately avoiding engagement with the empirical literature, or by oddly deferring to, and then co-opting, the empirical evidence alleged to provide the basis for some normative conclusion reached (Jalava and Griffiths 2017). Indeed, wading into the foreign waters of another discipline calls for epistemic humility whilst simultaneously disrupting the received view. But the brackish waters of neuroethics are strange to all disciplines that have preceded its slow individuation, though we humanitarians remain the stranger. This all the more underscores the need for interdisciplinary parity, for which I make a plea below.

Individuals diagnosed with Antisocial Personality Disorder (ASPD), or 'psychopaths' colloquially, are marked behaviorally by interpersonal violence, instrumental aggression, disregard for the rights of others, manipulativeness, deceitfulness,

---

nostic criterion appears to be a *sine qua non* for diagnosing any specific psychiatric disorder. For example, in DSM 5, criterion A for all personality disorders, including antisocial personality disorder, reads: "Significant impairments in self (self identity or self direction) and interpersonal (empathy and intimacy) *functioning*," [emphasis added].

[11] Credit to Adina Roskies for so identifying the two research programs in neuroethics.

persistent irresponsibility, and a lack of empathy.[12] This is a peculiar psychiatric disorder indeed, as it rather blatantly straddles the (alleged) divide between the moral and the medical. Hence the increasing interest taken by philosophers who often characterize psychopaths as 'morally blind agents' (e.g., Talbert 2008), that inflict violent harm onto others, and for whom we do not know how, or whether we even should attempt to, 'hold them responsible.' More recently philosophers have begun to make use of empirical research to advance arguments that psychopaths are not morally responsible agents and should not be blamed for their violent actions. Perhaps the most compelling such argument comes from neuroethicist Neil Levy (2005, 2007a, b). While we turn to Levy's argument, we need to keep firmly in mind Fulford's insight that evaluative content lies deeply hidden even within 'function,' the most fundamental concept within the naturalization cascade.

Levy (2005, 2007a, b) argues that psychopaths do not possess the capacity specifically for moral responsibility and that this lack of moral capacity is caused by a 'developmental disability,' for which there appears to be evidence of a dysfunctional neurological corollary (with specific reference to vmPFC function). Given the evidence, Levy argues, we should not hold psychopaths morally responsible. This is admittedly a mere cursory account of Levy's argument, but for current purposes I wish to examine the general structure of his argument, which fits quite perfectly into the general methodology in question that presumes the empirical facts (vmPFC function) are antecedently necessary for informing a normative conclusion (in this case, whether or not to engage in moral disapprobation of the psychopath).

The scenario appears to be that the normative moral question (whether or not to 'hold responsible') is deferential to the non-normative, empirical evidence (whether or not the psychopath 'is responsible'). Levy's view appears to assume that by 'is responsible' we are engaged in a query for which the content is non-normative and effectively reduces the normative query to a matter of vmPFC functioning, and that this non-normative matter-of-fact will simply wash out the normative upshot. This appears to be precisely what Levy (2007a, pp. 131–138; 2007b, pp.163–170) is arguing for: ASPD is a developmental disorder that is caused by neurological events that remain external to any blameworthy intra-psychological override (regardless of whether we name that capacity as a 'conscience,' or moral sense, or even moral knowledge) that we think we non-psychopaths possess. Levy argues that psychopaths and others with damage to the vmPFC, hence vmPFC dysfunction,

[12] *Diagnostic and Statistical Manual of Mental Disorders, Fifth Edition*. Also, to be clear, there is a mistake involved in simple equating those diagnosed with Antisocial Personality Disorder with those we call 'psychopaths.' Levy (2007) for instance, lumps the two together. These are indeed distinct constructions, the latter of which is thoroughly imbued with cultural presentations, but also within clinical discussions (e.g. Hare's Psychopathy Checklist; Hare 2003) in which a case is made that individuals scoring 23 points or higher are 'psychopaths' that are more entrenched in the symptoms than their lower-checklist-scoring Antisocial Personality Disordered counterparts. In fact, it might be too fast to merely see the 'psychopath' as a construction, for there is some question over whether or not psychopaths represent a distinct *taxon*, or a 'natural kind,' that is not merely constructed. I myself am not sympathetic to the argument that psychopaths represent a natural kind but rather a refinement of the more basic construction involved in ASPD.

undermines an agent's capacity for moral responsibility and renders those so afflicted not morally blameworthy for their violent actions. To be precise, Levy argues that if psychopaths never had a functioning vmPFC, unlike those with once functioning vmPFC's that were later damaged, then the former are categorically not morally responsible for having never developed the capacity (Levy 2007a, pp.130–131). This assumed the capacity for moral agency is in some essential way correlated with the functioning of the vmPFC. The correlation of vmPFC dysfunction to psychopathy, and otherwise morally questionable actions and attitudes of those with vmPFC damage is not drawn by me, but rather by Levy. So, considering the straw man objection that nobody believes there is a correlation between vmPFC and psychopathy, an available reply appears to be that Levy (2007a, b, p. 166), a preeminent figure in neuroethics, argues for such a correlation to the end that this correlation may well render psychopaths not morally responsible for their violence. Two objections lurk.

The first objection comes in the form of a Humean naturalistic fallacy charge that rejects the attempt to derive a normative 'ought' from a factual 'is.' The question over whether or not we can entail a normative 'ought' from an empirical 'is' without committing a fallacy takes new urgency in so-called applied ethics fields in which we are reliant on empirical evidence to provide some epistemic guidance to addressing our normative challenges.[13] We would be wrong to ignore empirical evidence relevant to normative consideration just as we would be wrong to uncritically accept a normative implication by overly relying on what appears to be even rather astounding empirical evidence. I do not here wish to wade into the general problem of the naturalistic fallacy and its newfound urgency as moral philosophy increasingly engages other disciplines but only register that the is/ought problem presents a serious fundamental objection to the entire enterprise of neuroethics. More importantly for current purposes is that a purported violation of the (alleged) naturalistic fallacy reveals the second objection, which is that what has likely occurred is that a norm has been smuggled into the antecedent 'is.' In order to get a norm out one must first put a norm in. And that is precisely what is occurring with respect to the 'function' of the vmPFC. Now, if 'function' were indeed normative in content, then Levy's position would be off the hook for the naturalistic fallacy charge because the factual status of 'is' is undermined, but this would come at the cost of giving up the epistemic merit bestowed on the empirical evidence. Levy (2007b) owns up to relying heavily on empirical studies of those with vmPFC damage to support his moral

[13] To be clear, I am here invoking the naturalistic fallacy to show that it can be averted but only at the price of ackowledging that there is no non-normative antecedent 'is' involved in neuroethics. Why? The 'function' of the vmPFC is itself normative in content, especially so when the given state of the vmPFC (i.e., fast/frugal or cold/calculated) is tied to a normative corollary. Now to the charge that invoking the naturalistic fallacy (which I am not simply doing) is currently out of fashion in philosophy. The customs of a given discipline are not in themselves an indication of good reasoning. Ignoring a fallacy does not make a fallacy disappear. Besides, if it is the case that many chapters in this volume raise problems for neuroethics by invoking the naturalistic fallacy, then the consensus that philosophers no longer care about it is also untrue. So here is another normative implication of neuroscience: the naturalistic fallacy should once again be taken seriously.

conclusion (that holding psychopaths responsible is wrong), and when called out specifically for this over-reliance (Nichols and Vargas 2008), doubles-down in response (Levy 2007b) by admitting his view to remain 'hostage to empirical fortune.' Any morality tied to such an empirically labile foundation in a nascent neuroscience yields nervous norms indeed!

At this point I wish to connect literature unrelated to the moral responsibility of psychopaths that converge on matters of neural function alleging to inform debate over the superiority of various normative moral theories (utilitarianism and deontology, specifically). I am aware of the risk of echoing much discussion in this volume surrounding the neuroscientific research of cognitive psychologists (Greene 2008; Cushman and Young 2008; Koenigs et al. 2007) and skeptics (Berker 2009), that correlates functional Magnetic Resonance Imaging (fMRI) results with neural activity related to normative moral replies to abstract, hypothetical moral dilemma thought experiments like the Trolley Problem and Footbridge Problem. I do so, however, for the dual purpose of showing that Levy's (2007a, b) conclusion, as well as Greene's conclusion (2008, that the evidence supports the normative superiority of utilitarianism over deontology), is undermined even at the empirical level (Koenigs et al. 2007). My attempt here is also to shine light on the hidden evaluative content of 'function' in so far as it is ascribed to the neuro-anatomical correlates of moral agency.

Very briefly summarizing Greene's (2008) compelling research seeking the underlying neural correlates of normative moral theory suggests that our brains essentially have two neural networks (one that is cold/calculated and another that is fast/frugal) that compete with one another (this is the 'dual hypothesis' thesis) and the 'cold' neural system correlates with utilitarian (welfare maximizing) responses to moral dilemmas whereas the 'fast' neural system correlates to characteristically deontological responses to moral dilemmas. From these neural correlations, so the contentious argument goes, we can see that utilitarianism, owing to its correlation with the 'cold' system, is the superior normative moral theory. Before raising objections to this argument below, I first want to address what appears to relatively straightforward counter-evidence that comes from Koenigs et al. (2007) that correlates *damage* to the vmPFC with increasingly utilitarian responses to moral dilemma scenarios. Recall Levy's empirically supported argument (2007) that psychopaths (like others with damaged vmPFC's) are entirely incapable of moral responsibility. Perhaps Greene was wrong and the 'cold' neural system as correlated with utilitarianism shows the theory to not be worthy of privilege after all, and for that matter the 'cold' neural system is possibly dysfunctional. This offsetting empirical evidence is troubling, yes, but it also highlights the bewitching nature of correlating the peculiar closeness of neuro-anatomical function with substantive moral verdicts.

Concerning the question of the normative content of 'function' with specific respect to neuro-function, we appear to have several options before us. We might either tie function to one leg (either the cold/calculated or the fast/frugal neural network) of the dual hypothesis thesis, or alternatively tie function to the vmPFC (or both insofar as the vmPFC is operative, or not, in either system). In order for 'function' to operate non-normatively in this case, we must identify – independently of normative presuppositions – the neural desiderata *in advance* of tying neuro-anatomy to any normative moral corollaries (by way of counterfactual moral dilemma scenarios), otherwise the neural desiderata themselves (fast/frugal, or cold/calculated, etc.) will simply reflect normative preference. But even having done so, there remains another lurking evaluative challenge to 'function' in that marking the neural desiderata themselves, independently of normative corollary considerations, requires a biological type of normative consideration that will always remain, to use Fulford's phrase, 'peculiarly logically close' to moral and social normative considerations.

8.4 Three Normative Implications of Neuroscience

First, the preceding analysis suggests that one normative implication of neuroscience is the supportive role neuroscience has in entrenching the biological/medical model of psychiatric disorder. Neuroscience appears to provide empirical basis in justifying the reconceptualization of a moral failing (vicious agent) as a pathological condition (psychopathy). In so far as neuroscience risks reinforcing the pathologization of heretofore-normative failings, it is decidedly participating in re-norming.

Second, given Fulford's compelling case that 'function' maintains normative content (and so for all concepts in the naturalization cascade), we need to rethink the correlations between anatomy and moral-dilemma verdicts as not at all a relationship between anatomical facts corresponding to specific moral verdicts. Rather, the neuro-anatomical functioning cannot be understood independently of the normative verdict we prefer from the outset. Not only would the attempt to derive neuro-anatomical ideals from normative verdicts be straightforwardly question-begging, doing so would so thoroughly imbue the neuro-anatomical ideal with normative content that yields good reason to doubt that there are any neuro-anatomical facts at all that are being correlated with normative verdicts. Strictly speaking, we would be correlating neuro-anatomical norms (based on desiderata of 'function') with moral norms.

Third, for all the attention neuroscience has bestowed to utilitarianism and deontology there has been relative inattention to the wider range of normative theories. Whence fMRI of the social contract theorist, varieties of feminist ethics, the natural law theorist, or the *eudaimon*? If only we could work backwards, equipped with the ideal neural profile of the moral saint, would we then be in a position of knowing what discrete regions of the brain signal in the right way, in the right circumstances, in the right time, and within the right person? This could only be achieved methodologically by a hypothetical life-long longitudinal, diachronic study of a single individual (that must also be ideally positioned in virtuous community). What would *that brain* look like, and from a diachronic perspective? As difficult as it may be to imagine this ideal brain it is not difficult to imagine that it would appear as something quite different than the current counterpart snapshots of brains that are responding to normatively limited abstract thought experiments. Alas, even if such an ideal moral profile were to be captured, it would be achieved at the cost of normative question begging.

8.5 A Plea for Interdisciplinary Parity: Toward Neurohumanism

We need disciplinary resources for the neuro*humanities*, of which neuroethics comprises merely one part, in order to balance the power of the neuroscience in the marketplace of ideas. If one takes seriously Thomas Kuhn's insight (Kuhn 1962) into the sociological forces of scientific theory advancement that beyond more or new empirical evidence, theoretical paradigm shifts require new interpretations of the evidence as well as preponderance of acceptance of a given interpretation by the intellectual community - then there is all the more reason to alter the intellectual community by way of interdisciplinary plurality.[14] Doing so might well slow the interpretations of advancements in neuroscientific research by way of injecting far more skepticism about the normative inferences and conclusions reached by those active in neuroscientific research. The sheer volume of product emerging from neuroscientific research raises the suspicion that this nascent yet well-funded field is epistemically unrestrained. Besides, why think neuroscientists and cognitive psychologists ought to remain on the interpretive pedestal? The perspectival plurality afforded by an interdisciplinary approach to neuroscientific studies will better position the intellectual community to face the normative complexities inherent to the human condition. This can only be accomplished with lateral interdisciplinarity. No pedestal is required.

If we are not careful in checking our epistemic ambitions surrounding neuroscientific insights, then we run the substantial risk of burying normative assessment under waves of neuroscientific data.

References

American Psychological Association. 2013. *Diagnostic and Statistical Manual of Mental Disorders*. 5th ed. Washington, DC.

Berker, S. 2009. The Normative Insignificance of Neuroscience. *Philosophy & Public Affairs* 37 (4) Wiley Periodicals.

Boorse, C. 1975. On the Distinction Between Disease and Illness. *Philosophy & Public Affairs* 5: 49–68.

———. 1997. A Rebuttal on Health. In *What Is Disease?* ed. Ed.J. Hunter and R. Almeder, 1–134. Totowa, NJ: Humana Press.

[14] See Kuhn (1962): *The Structure of Scientific Revolutions.*

Callahan, D. 1973. World Health Organization Definition of Health. *Hastings Center Studies* 1 (3) The Hastings Center.

Charland, L.C. 2004. Moral Treatment and the Personality Disorders. In *The Philosophy of Psychiatry: A Companion*, ed. Jennifer Radden. Oxford: OUP.

Cleckley, H.C. 1955. *The Mask of Sanity*. St. Louis: The C.V. Mosby Company.

Cushman, F., and L. Young. 2009. The Psychology of Dilemmas and the Philosophy of Morality. *Ethical Theory Moral Practice* 12: 9–24. Springer.

Daniels, N. 2007. *Just Health: Meeting Health Needs Fairly*. Cambridge: Cambridge University Press.

Elliot, C. 1996. *The Rules of Insanity: Moral Responsibility and the Mentally Ill Offender*. Albany: State University of New York Press.

———. 2004. Mental Illness and Its Limits. In *The Philosophy of Psychiatry: A Companion*, ed. Jennifer Radden. Oxford: OUP.

———. 2010. *White Coat Black Hat: Adventures on the Dark Side of Medicine*. Beacon Press.

Fulford, K.W.M. 1989. *Moral Theory and Medical Practice*. OUP.

———. 1994. Not More Medical Ethics! In *Medicine and Moral Reasoning*, ed. Fulford, Gillett, and Soskice. Cambridge University Press.

———. 2000. Teleology Without Tears: Naturalism, Neo-naturalism, and Evaluationism in the Analysis of Function Statements in Biology (and a Bet on the Twenty First Century). *Philosophy, Psychiatry and Psychology* 7 (1): 45–65. Johns Hopkins University Press.

———. 2004. Facts/Values: Ten Principles of Values-Based Medicine. In *The Philosophy of Psychiatry: A Companion*, ed. Jennifer Radden. Oxford: OUP.

Fulford, K.W.M., M. Davies, R.T. Gipps, G. Graham, J.Z. Sadler, G. Stanghellini, and T. Thornton. 2013. *The Oxford Handbook of Philosophy and Psychiatry*. Oxford: OUP.

Garnar, A., and V.G. Hardcastle. 2004. Neurobiological Models. In *The Philosophy of Psychiatry: A Companion*, ed. Jennifer Radden. Oxford: OUP.

Greene, J. 2008. The Secret Joke of Kant's Soul. In *Moral Psychology, Vol.3: The Neuroscience of Morality: Emotion, Brain Disorders and Development*, ed. Walter Sinnott Armstrong, 35–79. Cambridge, MA: MIT Press.

Hacking, I. 1995. *Rewriting the Soul: Multiple Personality and the Sciences of Memory*. Princeton: Princeton University Press.

———. 1998. *Mad Travelers: Reflections on the Reality of Transient Mental Illness*. Charlottesville: University Press of Virginia.

Hare, R.D. 2003. *The Psychopathy Checklist*. Toronto: Multi-Health Systems. Revised second edition.

Jalava, J., and S. Griffiths. 2017. Philosophers on Psychopaths: A Cautionary Tale in Interdisciplinarity. *Philosophy, Psychiatry & Psychology* 4 (1) Johns Hopkins Press.

Kendall, R.E. 1975. The Concept of Disease and Its Implications for Psychiatry. *British Journal of Psychiatry* 127: 305–315.

Koenigs, M., L. Young, R. Adolphs, D. Tranel, F.A. Cushman, and M.D. Hauser. 2007. Damage to the Prefrontal Cortex Increases Utilitarian Moral Judgments. *Nature* 446: 908–911.

Kuhn, T. 1962. *The Structure of Scientific Revolutions*. Chicago: Chicago University Press.

Levy, N. (2005) The Good, The Bad and the Blameworthy. *Journal of Ethics Social Philosophy* 1 (2): 1–16

———. 2007a. The Responsibility of Psychopaths Revisited. *Philosophy, Psychiatry and Psychology* 14 (2): 129–138. Johns Hopkins University Press.

———. 2007b. Norms, Conventions, and Psychopaths. *Philosophy, Psychiatry and Psychology* 14 (2): 163–170. Johns Hopkins University Press.

Nelkin, D. 2015. Psychopaths, Incorrigible Racists, and the Faces of Responsibility. *Ethics*: 357–390.

Shaun, Nichols, Vargas, Manuel. 2008. *How to Be Fair to Psychopaths. Philosophy, Psychiatry, & Psychology* 14 (2): 153–155.

Nordenfelt, L. 1987. On the Nature of Health: An Action-Theoretic Approach. In *Philosophy and Medicine*, vol. 26. Dordrecht, Holland: D. Reidel Publishing.

Radden, J. 2004. In *The Philosophy of Psychiatry: A Companion*, ed. Jennifer Radden. Oxford: OUP.

Talbert, M. 2008. Blame and Responsiveness to Moral Reasons: Are Psychopaths Blameworthy? *Pacific Philosophical Quarterly* 89: 516–535.

———. 2011. Unwitting Behavior and Responsibility. *Journal of Moral Philosophy* 8: 139–152.

Tan, J., A. Stewart, R. Fitzpatrick, and T. Hope. 2007. Competence to Make Treatment Decisions in Anorexia Nervosa: Thinking Processes and Values. *Philosophy, Psychiatry and Psychology* 13 (4) Johns Hopkins University Press.

Tekin, Serife. 2017. Are Mental Disorders Natural Kinds? A Plea for a New Approach to Intervention in Psychiatry. *Philosophy, Psychiatry and Psychology* 23 (2): 147–163. Johns Hopkins University Press.

Thornton, T. 2000. Mental Illness and Reductionism: Can Functions Be Naturalized? *Philosophy, Psychiatry and Psychology* 7 (1) Johns Hopkins University Press.

Wakefield, J.C. 1992. The Concept of Mental Disorder: On the Boundary Between Biological Facts and Social Values. *American Psychologist* 47 (3): 373–388. Johns Hopkins University Press.

———. 1995. Dysfunction as a Value-Free Concept. *Philosophy, Psychiatry & Psychology* 2 (3): 233–246. Johns Hopkins University Press.

———. 2000. Aristotle as Sociobiologist: The Function of a Human Being Argument, Black-Box Essentialism, and the Concept of Mental Disorder. *Philosophy, Psychiatry and Psychology* 7 (1): 17–44. Johns Hopkins University Press.

———. 2009. Mental Disorders and Moral Responsibility: Disorders of Personhood as Harmful Dysfunctions. *Philosophy, Psychiatry & Psychology* 16 (1): 91–99. Johns Hopkins University Press.

Watson, Gary. 1993. Responsibility and the Limits of Evil: Variations on a Strawsonian Theme. In *Perspectives on Moral Responsibility*, ed. John Martin Fischer and Mark Ravizza. Ithaca: Cornell University Press.

———. 1996. Two Faces of Responsibility. *Philosophical Topics* 24 (2).

Chapter 9
Neuromodulation of the "Moral Brain" – Evaluating Bridges Between Neural Foundations of Moral Capacities and Normative Aims of the Intervention

Christian Ineichen and Markus Christen

Abstract The question of whether neuroscience has normative implications or not becomes practically relevant when neuromodulation technologies are used with the aim of pursuing normative goals. The historical burden of such an endeavor is grave and the current knowledge of the neural foundations of moral capacities is surely insufficient for tailored interventions. Nevertheless, invasive and non-invasive neuromodulation techniques are increasingly used to address complex health disturbances and are even discussed for enhancement purposes, whereas both aims entail normative objectives. Taking this observation as an initial position, our contribution will pursue three aims. First, we summarize the potential of neuromodulation techniques for intervening into the "moral brain" using deep brain stimulation as a paradigmatic case and show how neurointerventions are changing our concepts of agency and personality by providing a clearer picture on how humans function. Second, we sketch the "standard model" explanations with respect to ethically justifying such interventions, which rely on a clear separation between normative considerations ("setting the goals of the intervention" or "the desired condition") and empirical assessments ("evaluating the outcome of the intervention" or "the actual condition"). We then analyze several arguments that challenge this "standard model" and provide bridges between the empirical and normative perspective. We close with the observation that maintaining an analytical distinction between the normative and empirical perspective is reasonable, but that the practical handling of

C. Ineichen
Department of Psychiatry, Psychotherapy and Psychosomatics & Department of Neurology, Psychiatric & University Hospital Zurich, Zürich, Switzerland
e-mail: christian.ineichen@uzh.ch

M. Christen (✉)
Institute of Biomedical Ethics and History of Medicine, University of Zurich, Zürich, Switzerland
e-mail: christen@ethik.uzh.ch

G. S. Holtzman, E. Hildt (eds.), *Does Neuroscience Have Normative Implications?*, The International Library of Ethics, Law and Technology 22,
https://doi.org/10.1007/978-3-030-56134-5_9

neuromodulation techniques that involve normative intervention goals is likely to push such theoretical distinctions to their limits.

Keywords Neuromodulation · Deep brain stimulation · Is-ought-gap · Agency · Personality · Self-regulation

9.1 Introduction

The potential impact of the natural sciences on normative inquiries has been the basis of considerable debate. After the collapse of the Thomistic worldview supporting the view that the good, the true and the being are three perspectives of the same reality (e.g. Rüfner 1964), the scientific and the moral domain have been carefully separated. It has been a rather recent development that we now again discuss interconnections of these two domains (see e.g. Harris 2011). That neuroscience started to infiltrate "morality" as a research area by investigating the neural foundations of human moral behavior at least since the tragic case of Phineas Gage, a railroad worker whose skull was brutally perforated by an iron rod during an accidental explosion, can be considered one reason for this rapprochement. Gage, who showed sudden asocial behavioral manifestations secondary to the accidental explosion, set the stage for scientifically inquiring into the "moral" brain. And although a closer look at the story revealed that many of the accounts of Gage's life after 1848 are strange mixtures of slight fact, considerable fancy and downright fabrication (Macmillan 2000), Gage symbolized a paradigmatic description of how brain and moral behavior are related – namely in the sense that the dysfunction of some parts of the brain (the right orbitofrontal or ventromedial prefrontal cortex) can lead to major aberrations in moral behavior.

Interestingly, this "deterministic connection" of lesions in certain brain regions with amoral behavior has already been questioned in the first "review paper" (using today's terminology) on the effect of frontal lesions on behavior by Leonore Welt, published in 1888 (2009). Welt presented 58 cases (of various forms) of frontal lesion patients (including the Phineas Gage case), of whom 47 did not show character changes ("Charakterveränderungen") after the lesions (Christen and Regard 2012). Certainly, degree and localization of these injuries were much harder to describe when neuroimaging was not yet available. Nevertheless, Welt urged caution for those seeking to deterministically associate brain lesions with character changes. Irrespective of this, the case of Phineas Gage is frequently cited in the introduction of papers published today that discuss the relationship between brain and morality.

This emerging knowledge about the connection between brain and (moral) behavior also triggered the tragic period between 1935 and 1950, characterized by the dramatic rise and subsequent crashing fall of psychosurgery to treat mental illness. The neurophysiologist and co-recipient of the Nobel Prize in Medicine in 1949, Egaz Moniz, first proposed in 1935 to treat a patient's pathological anxiety states and paranoid ideas by disrupting the connections between the frontal lobes and the rest of the brain, a procedure called "prefrontal leucotomy" (Valenstein 1986)—a term that was later changed to lobotomy by Walter Freeman. With only anecdotal information to support his theory, no evidence from preclinical experimentation, no tests of safety or effectiveness and without any follow-up of patients beyond the immediate postoperative period, Moniz saw his intervention quickly adopted by several other medical practitioners. In particular, Walter Freeman almost single-handedly popularized the widespread use of psychosurgery throughout the U.S. between his first prefrontal lobotomy performed on the 14th September in 1936 and his last lobotomy in 1967 by performing about 3400 transorbital lobotomies (Valenstein 1986; El-Hai 2005). Despite critics describing lobotomized patients to be confused, lacking in affect and without normal response to common human interactions, there was a remarkable lack of critical evaluation by rapidly declaring many of the patients treated as completely cured. The tragic rise and belated fall of psychosurgery in the 1970s elucidates how science attempted to pursue normative goals by modulating people's brain, aiming at "recalibrating" their behavior and affect so as to conform to certain societal norms.

Meanwhile, the more recent advent of trying to empirically tackle the foundations of human moral behavior has snowballed into numerous inquiries that are linked to the promise of providing solid scientific answers not only to questions of moral behavior, but also to questions of morality as such. Contributing to this emerging aura of authority has been the close collaboration of researchers from psychology, philosophy and neuroscience, with the aim of illuminating the distinct neuronal pathways that underlie our moral intuitions, motivations, judgments and behaviors. For example, Joshua Greene, a philosopher by training, joined forces with neuroscientist Jonathan Cohen and analyzed functional magnetic resonance images of people's brains as they responded to hypothetical moral dilemmas (Greene et al. 2001). Another example includes social psychologist Jonathan Haidt, who, based on a set of neuropsychological experiments, questioned the role reasoning plays in moral judgments (Haidt 2007). In his view, reason rather serves only a secondary role as a post-hoc means of justifying unreflective and quick intuitions. Moreover, many philosophers (beginning with Foot (1967) and Thomson (1985)) and experimental psychologists (starting with Greene et al. 2001) have been especially taken with a series of moral conundrums referred to as "trolleyology" – a provoking series of thought experiments circling around the question of whether one should (intentionally or accidentally) sacrifice a human being to prevent that a runaway trolley kills several human beings (Cathcart 2013). For this empirical inquiry, numerous techniques, including neuroimaging, have been endeavored.

The story above, about neuroscientific investigations and their ability to provide stringent answers to questions of morality, would be one-sided if one would not mention that this scientific advance ignited a countermovement, typified by e.g. philosopher Selim Berker. Berker has pointed out in his paper "the normative insignificance of neuroscience" the claim that neural and psychological processes that track personal factors cannot be relied on to support moral propositions or guide moral decisions (Berker 2009). Consistently, others have convincingly claimed that an understanding of the neural correlates of reasoning cannot tell anything significant about whether the outcome of this reasoning is justified. The discussion above already touches on unjustifiably bridging the is-ought gap, that is, suggesting that knowledge gained through descriptive natural science can itself determine what is morally right and wrong.

The previously sketched focus of neuroscientific studies investigating the neural foundations of morality has led to a first and incomplete mapping of the "moral brain". Meanwhile, a large body of research has tried to identify and probe "moral circuits" that underlie human processing both in healthy and pathological states. Consequently, various brain regions have been implicated in emotional processing and social cognition, including theory of mind. Those include, among others, the ventromedial prefrontal cortex (including the medial orbitofrontal cortex and Brodmann area 10 and 11), the orbitofrontal cortex, but also the amygdala, superior temporal sulcus, bilateral temporoparietal junction, posterior cingulate cortex, and precuneus (Young and Dungan 2012; Fumagalli and Priori 2012; Dolan 1999; Casebeer 2003). However, and even though many brain areas have successfully been associated with moral motivation, judgment and behavior, it is important to highlight that insight on the interconnection between moral abilities and brain regions is still rather scarce and unspecific, as the associated networks and brain regions are involved in a broad range of human behaviors.

What is more relevant in the context of the neuroscience of morality is that advances in contemporary neuroscience have made it possible to probe and modulate brain functions with an unprecedented level of precision and with hypotheses gleaned from preclinical (using laboratory animals) and clinical research. Some of the most recent neuromodulation technologies have made it possible to probe causal relationships between the different dysfunctional socio-moral pathways underlying many complex (e.g. neuropsychiatric) pathologies. Examples of such technologies include deep brain stimulation (DBS), transcranial magnetic stimulation (TMS), transcranial direct current stimulation (tDCS), and optogenetics, to name a few. All these technologies manipulate neural networks in a rather specific, evidence-driven way. While the level of precision with which neural elements can be modulated varies between the different approaches, they all surely transcend the intervention possibilities commonly employed by oral medication. Network science, as an approach to model and analyze the dynamics of communication on networks, has proven useful for the prediction of emergent network states. Hence, such approaches can offer insight into the mechanisms by which brain networks transform and process information (Avena-Koenigsberger et al. 2018). The aim of many researchers of injecting test signals into an electrical neural circuit for subsequent recording of the output at distinct points also constitutes an engineering ideal. This portrays nicely the image of a more technical notion of the human brain.

The question of whether neuroscience has normative implications or not becomes relevant when neuromodulation technologies are used with the aim of pursuing normative goals. As outlined previously in the discussion of psychosurgery, the historical burden of such an endeavor is grave and the current knowledge of the neural foundations of moral capacities is surely insufficient for tailored interventions. Nevertheless, invasive and non-invasive neuromodulation techniques are increasingly used to address complex health disturbances and are even discussed for enhancement purposes (Earp et al. 2017), and both of these aims entail normative goals. In what follows, we describe DBS, a neuromodulation technology promising much hope in the treatment of numerous complex health disturbances, and its potential of intervening into the "moral brain". As will be seen, research on DBS has provided insights into the neurophysiology of the basal ganglia and how a specific area of the brain, namely the subthalamic nucleus (STN), is involved in various computations necessary to establish agentic states. Apart from agency, DBS-research has been associated with complex behavioral alterations that encompass personality-constituting processes. After describing complex psychological alterations following DBS interventions, we provide a theoretic link that tries to bring together notions of agency and self-regulation on the ground of the neurophysiology of the basal ganglia. We then outline how both notions have relevance in the moral sphere. Finally, we provide normative aspects of neuromoral interventions that will focus on the presumed separability of the normative and empirical perspective that have practical relevance for medical decision-making processes.

9.2 Neuromodulating the "Socio-moral Brain"

In what will follow, we shall turn towards deep brain stimulation as an exemplary case for intervening into complex health disturbances. We will extend this notion by alluding to complex disorders that inevitably involve complex psychological aberrations with social and moral relevance.

9.2.1 DBS as an Exemplary Case of Intervening Into the "Moral Brain"

It was 1971 when psychologist B. F. Skinner, the father of behaviorism, expressed his hope that some of the most abominable of the humanly created problems that derange our lives, such as wars and famines, could all be solved by new "technologies of behavior." Skinner wanted to "maximize the achievements of which the human organism is capable" – an aim that already subsumes normative goals pertaining to social engineering (Skinner 1972). Albeit Skinner understood these "technologies of behavior" as manipulation of the external environment at that time, recent advances have emerged to the point where manipulations of the internal environment of the brain are possible. One such possibility is exemplified by the electrical modulation of brain circuits through DBS, a symptomatic neurosurgical intervention (i.e. leading to substantial symptom relief in well-selected patients) that includes electrode implantation to apply electrical currents to target structures.

It was 1976 when Jean Siegfried started DBS in pain patients and 1986, at the meeting of the American Society for Stereotactic and Functional Neurosurgery, when Alim-Louis Benabid presented his abstract on stimulation of the ventral intermediate nucleus of the thalamus (Vim) for Parkinson's disease (PD). Both events marked the beginning of modern deep brain stimulation (DBS) (Christen and Müller 2012; Hariz 2012). From this moment on, intense research activities marked an era of probing new anatomical targets, resulting in a subsequent increase in the therapeutic spectrum. The early success of approved and exempted indications has led to the use of DBS to treat a non-undisputed variety of other emerging indications, including e.g. pain, obsessive compulsive disorder, major depression, Tourette syndrome, obesity, anorexia nervosa, substance addiction, epilepsy, pathological aggression, dementia, apart from different movement disorders (Youngerman et al. 2016). In particular, the early success and the subsequent research focus on treating PD through DBS led to multiple randomized clinical trials that have shown DBS to be more effective than the best medical treatment in well-selected patients (Deuschl et al. 2006; Follett et al. 2010; Weaver et al. 2009). The recognition that DBS was better at improving the motor symptoms of PD compared to best medical treatment was final evidence for DBS to be taken as the gold standard therapeutic intervention for advanced PD that can no longer be successfully treated with L-Dopa.

Meanwhile, the ongoing expansion of DBS indications from movement disorders such as dystonia, essential tremor and PD to targeting mood and mind (in e.g. depression and Alzheimer's disease) marked a time that was characterized not only by intense research activities but also by serendipitous discoveries. For example, a failed attempt to treat obesity in a patient showing vivid memory recollection during stimulation (Hamani et al. 2008) expeditiously resulted in a first clinical trial for Alzheimer's disease (Laxton et al. 2010). As a result of intended and incidental discoveries, DBS is now used for treating various indications reaching from cognition to volition to behavior and affect. Owing to improvements in navigation and imaging techniques, researchers were further able to pursue hypothesis driven empirical research. In parallel, often observed side-effects built the basis for a broadening of the therapeutic spectrum. Some of the observed side effects encompassed complex affective and behavioral changes relating to personality that are more difficult to assess and to describe. Unsurprisingly, the observation and description of these complex alterations has taken longer compared to more focally described changes such as parasthesias (abnormal sensation such as tingling, numbness) or disconjugate gaze (unpaired movements of the eyes).

Currently, DBS is still most frequently used for treating patients suffering from PD. PD represents a complex, chronic, neurodegenerative, multi-systemic disorder that not only affects the dopaminergic system but also serotonergic, glutamatergic and GABAergic neurotransmission (Tremblay et al. 2015), leading to motor and non-motor dysfunction (Schapira et al. 2017). Accordingly, PD additively involves non-motor symptoms such as neuropsychiatric disturbances (depression, anxiety, apathy), sleep-wake dysfunction (e.g. REM sleep behavior disorder and excessive daytime sleepiness), cognitive problems (e.g. dysexecutive syndrome, dementia) and autonomic dysfunctions (e.g. sexual dysfunction, orthostasis) (Olanow et al. 2011).

Mounting evidence suggests that DBS interventions, primarily geared to treating PD, can result in complex behavioral and affective changes such as hypomania, new onset impulse control disorders (ICDs), including hypersexuality, pathological buying, pathological gambling, and addiction to levodopa (Ballanger et al. 2009; Hack et al. 2014; Castrioto et al. 2014), logorrhea, irritability, aggression, mirthful laughter, egocentrism, lying, and acute sadness and crying (summarized in Ineichen et al. 2016; but see also: Jahanshahi et al. 2015; Müller and Christen 2011). Currently, it is difficult to pinpoint whether such complex non-motor side effects emerge from stimulation, ongoing pathological processes of the disorder being treated, pharmacological therapy, or their interrelation (Gilbert et al. 2018, 2020). Nevertheless, through sophisticated study designs, such neuromodulation interventions have the potential to provide causal explanations by demonstrating whether modulation of a certain node is necessary and sufficient for generating the hypothesized modification in function. Therefore, just as the description of side effects that paved the way for the exploration of DBS in areas beyond movement disorders, there is a possibility that these DBS-related complex side effects could provide a basis for future interventions that target socio-moral abilities of patients. In 2012 for example, the journal *Brain* published an article titled "Functional and Clinical Neuroanatomy of Morality" in which the authors allude to the functional role of subcortical structures in the context of morality (Fumagalli and Priori 2012). They also suggest that DBS might be used to treat pathological antisocial behaviors.

Evidently, the above-mentioned alterations following DBS interventions pertain to personality-related changes comprising changes in socio-moral information processing. Given that the treatment of patients suffering from different disorders with electrodes implanted in various nuclei of the brain resulted in a more refined understanding of computations underlying various neural processes, the same is likely to take place in terms of understanding neural computations that underlie socio-moral behavior. As knowledge increases, it is conceivable that similar practices could take place with diseases that directly affect moral behavior, and such engaging in such practices could be the first step toward a more general understanding of the interaction between processes in the brain and moral behavior. Although we have to acknowledge that current DBS interventions interfere with complex socio-moral networks (illuminating the present impossibility to deliver current to target in an appropriate and specific way; Ineichen et al. 2018), they also demonstrate the potential to modulate our social and moral infrastructure through invasive brain technologies. Even though e.g. neuroimaging has provided much insight into neural processes, simply building strong correlations through sophisticated modeling and data analysis still does not provide causal mechanistic insights (Poldrack and Farah 2015). Because of their potential to probe brain networks, neuromodulation technologies, such as DBS, are gaining an increased amount of attraction.

9.2.2 Agency, Control and Self-Regulation

Because neuroscience has just started investigating the mechanisms that generate agency, personality and complex socio-moral computations in general, and moral behavior specifically, there is not yet precise knowledge of how to intervene into these capacities. Nevertheless, we can succumb to the idea that it might only be a matter of time until precise knowledge on this front will surface. Hence, in what follows let us assume that there will be precise knowledge of the neural infrastructure of moral capacities available in the near future, and that brain technologies could in principle be used to provide tailored interventions that aim at altering human moral capacities. For in fact, making humans more "ethical" by synthesizing knowledge on the neural basis of moral decision-making or the neural infrastructure of self-regulation really might soon become a realistic goal. Notwithstanding its logical content, the assumption of altering moral capacities may appear less ecologically invalid if some delineations are made regarding the human capacity to self-regulate, to exert control and to be an agent of one's life. Self-regulation moreover represents a paradigmatic case of a moral capacity itself. As will be seen, DBS of the often-targeted basal ganglia, e.g. in the context of PD, can play an important role with respect to such a capacity.

9.2.2.1 Agency & Self-regulation: Concepts and Relation to Morality

In the sense of the term which we use here, "agency" refers to the ability to translate mental states such as desires, beliefs or intentions into executive plans of action and encompasses a set of sensorimotor, cognitive, affective and volitional components. Neurological and neuropsychiatric disorders impose internal constraints on a number of these components and hamper the ability to act. While PD can partly be characterized as a hypokinetic movement disorder with patients showing difficulty to execute movements because dysfunction of the basal ganglia undermines primarily the sensorimotor component of agency, major depression and generalized anxiety, for example, impair the cognitive, affective and volitional components of agency. Compared to hyperagentic disorders that show an excessive experience of one's own causation and control over events, as evidenced in obsessive compulsive disorder (OCD) and schizophrenia, PD and major depression can be termed "hypoagentic" disorders leading to loss of control and an inability to carry out action plans (Ineichen and Christen 2017; Glannon and Ineichen 2016; Haggard 2017). DBS can restore some degree of agency in some of these conditions by allowing executive control. As will be seen, agency can be considered a prerequisite for self-regulation because any pathophysiological process that undermines the patients' ability to convert volitions, thoughts and emotions into actions will typically impede some forms of self-regulation. Insofar as autonomy can include competency and authenticity (Mele 2001) and as such represents an important personality trait, various neurological and psychiatric states can impede patients' abilities to critically reflect, carry out, identify with or endorse certain mental states (Glannon and Ineichen 2016).

The theme of willpower, primarily focusing on cognitive and volitional subsets of agency at that time, features strongly in ancient mythical stories reaching as far back as Homer's Odyssey. Self-control also emerged as a unifying theme during the Victorian era in the nineteenth century when the building of character and the exercise of willpower were considered synonymous with one another. In fact, self-regulation is an important topic in evolutionary anthropology and also in models trying to describe the genealogy of morality. Apart from the notion that the capability of inhibiting certain instincts might be a true *specificum humanum*, it is adaptive to possess the ability not only to give room for the stronger instinct (e.g. in cases of competing goals) but to be able to consider prospect of success and the like. Sometimes, it might be better to wait for a while and to follow another, yet less important goal in the meanwhile. In the tradition of Arnold Gehlen, an important figure in philosophical anthropology, this phenomenon was termed "the hiatus of self-control" (Gehlen 1940). Being able to inhibit certain goals for the sake of other executive plans—possibly only experienced in a modus of anticipation—implies the development of a complex sense of time including anticipation and foresight. In fact, not only future actions have to be anticipated but also the motivational moods and motives, which help coordinating the respective action. As a result, executive control can be conceptualized as a defining characteristic of morality, as putting aside own desires is required if perceiving that intentions and goals of others might be harmed. Also Wallace (Wallace 1994) articulated that free autonomous agents must possess the powers of "reflective self-control". The idea, that self-regulation involves the capacity to stand back from and evaluate one's first-order desires is furthermore widely shared by philosophers of both a Humean and a Kantian cast (e.g. Bratman 2000 or Kennett 2001).

9.2.2.2 Current Understanding of Self-regulation in a Moral Context

Similarly, today the capacity of the human mind to alter its own responses is regarded as an important foundation for culture, achievement, individual success and morality. The ability to suppress or override competing responses and to alter one's responses so as to bring them into line with individual goals, ideals, moral values, social norms, laws, and other standards is an important personality process (Kahneman and Chajczyk 1983; Jonides and Nee 2006; Tangney et al. 2004; Gross 1998). It is therefore a vital foundation for the building of character and a core feature of human agency (Baumeister et al. 2006; Bandura et al. 2003). Self-regulation strongly connects to morality insofar as it refers to the "capacity to exercise moral management of the temptations and provocations the individual encounters in a setting" (Wikström 2005, p. 217). Further, people with stronger self-regulation capabilities are expected to be less likely to rely on conscious active self-control because the more energy conserving implicit self-control mechanisms are already robustly installed and therefore are more likely to behave morally.

9.2.2.3 Basal Ganglia and the Neurophysiology of Self-regulation

Meanwhile, a number of brain areas have been implicated in cognitive control, conflict monitoring, response selection and -inhibition and self-regulation in general – among others, the prefrontal cortex (PFC) (Petrides and Pandya 1999), the basal ganglia (Nambu 2008) and the anterior cingulate cortex (ACC) (e.g. Somerville et al. 2006). Low levels of self-control have also been associated with a less refined connectivity (e.g. less myelination and orientation regularity) in frontostriatal (Liston et al. 2006; Casey et al. 2007) and frontoparietal circuitry (Jonides et al. 2000), critical for effective cognitive control. In the classic basal ganglia model, the output nuclei (globus pallidus internus and substantia nigra pars reticulata) hold the cortex and superior colliculus under tonic inhibition to prevent inappropriate (i.e. maladaptive) movements and can phasically release this inhibitory control to allow movements if medium spiny neurons are in the physiological upstate and enough dopamine is present (Jahanshahi et al. 2015; Da Cunha et al. 2015). Executive control is important to override habitual or prepotent responses, to control emotions, focus attention and exert self-regulation necessary for social interaction. Mounting evidence suggests that the basal ganglia coordinate behavioral output through neuronal inhibition and disinhibition, thereby enabling flexible interactions and executive control relevant in cognitive and emotional processing. Therefore, adaptive behavior owes as much to taking appropriate action as to inhibiting or suppressing contextually inappropriate or socially unacceptable behavior (Jahanshahi et al. 2015). Consistently, failure of context-appropriate inhibitory control through e.g. frontostriatal dysfunction leads to many impulsive manifestations relevant in neuropsychiatric disorders. The STN of the basal ganglia, the physiological function of which is to raise the response threshold temporarily and to provide time for information accumulation before decision making and responding, has important connections to the ventral tegmental area and ventral striatum, implicated in drug craving, emotion regulation and impulsivity (Kober et al. 2010; Frank et al. 2007). In fact, the STN is nowadays generally considered to be not only a regulator of motor function, but also an important gateway for cognitive and emotional signal processing. A recent review including tracing, cytoarchitectonic, imaging and electrophysiologic studies furthermore points toward a topographic segregation (motor, associative and limbic) within the STN (Lambert et al. 2015; Temel et al. 2005). It is therefore possible that interference with the STN by disease (PD) or intervention (drugs, DBS), can modulate associative and limbic processing (Tremblay et al. 2015; Kopell and Greenberg 2008; Sesack and Grace 2010). As has been described, STN-DBS, albeit primarily intended to treat motor symptoms of PD, also modulates associative and limbic basal ganglia networks with behavioral and psychological consequences. Complex alterations including personality changes postoperatively may in fact point to aberrant self-regulation that might have been overshadowed by a generalized lack of agency. Studies corroborate this hypothesis insofar as they describe e.g. new onset ICDs following DBS. In sum, stimulation of the basal ganglia nuclei through DBS could alter one's ability to self-regulate as the applied current intervenes into the neural basis of emotional, motor and cognitive self-regulation. This exemplifies the potential of DBS to intervene into the complex infrastructure of the human brain with the aim of altering moral capacities including e.g. self-regulatory abilities.

9.3 Normative Considerations of Interventions into the "Moral Brain"

Characteristic of any medical intervention is the need to conduct a harm-benefit[1] analysis. Typically, harm-benefit analyses are performed in the pre-intervention phase with the aim of determining whether the upcoming procedure is justified. In turn, it is used as a basis to retrospectively assess the success of the therapy for the individual patient. In addition, a post-interventional assessment is often used to collect data on the effectiveness and risk of the performed procedure for future patients. Generally, this involves a thorough analysis of the aims for which the intervention is applied followed by rigorously examining whether these aims have been fulfilled. This widely-adopted process also addresses the question of whether the therapeutic aim is at all desirable and therapeutically reachable.

Such a comprehensive ethical harm-benefit assessment certainly also holds for interventions with the aim to change moral behavior. While we have spent some time to substantiate that next generation interventions might soon be installed for changing moral abilities of patients, the ethical question of the desirability of the intervention still needs clarification. As can be anticipated, the pre- and post-interventional harm-benefit analysis will be different depending on whether neuromoral interventions are at stake. Consistent with the argument presented above, it is important to stress that understanding neurophysiological processes does not replace the evaluation of outcomes. In what follows, we first outline what we call the "standard model" for ethically justifying interventions, and then question the perceived inseparability of intervention aims and outcomes if moral capacities become part of such intervention practices.

[1] We deliberately use the expression "harm-benefit analysis" instead of the more commonly used term "risk-benefit" because the latter is misleading.

9.3.1 Roots of the "Standard Model"

What we call the "standard model" of any ethical evaluation of a medical intervention is the clear-cut separation between the definition of (and agreement upon) the intervention goals and the objective verification of whether these goals have been met. The former is usually seen as a normative endeavor, whereas the latter is considered an empirical task.

This "standard model" reflects a basic analytic distinction in ethics – namely the distinction between an "is" (the actual condition; measured in the evaluation) and an "ought" (the desired condition, determined as the therapeutic goal). This requirement of a clear-cut distinction between "is" and "ought" is often referred to in the work of David Hume. In his 1740 published "Treatise of Human Nature" he pointed to the importance to distinguish between indicative or descriptive and imperative or prescriptive statements and that one cannot justify normative claims solely by relying on empirical (descriptive) premises. Furthermore, in contrast to the inspection of the truth-value of descriptive statements, prescriptive statements are not intended to provide verification or falsification, but instead have the character of a legitimation. Further, Hume attests that verification alone can never lead to a legitimation. Accordingly, the is-ought-fallacy is committed if one derives an "ought" from an "is". Of course, there is a reciprocal relationship in that one is equally wrong when deriving an "is" from an "ought" (termed the moralistic fallacy). To avoid falling into this trap, a clear separation between normative and empirical claims is advocated and most people defend the idea of not confusing these two, presumably separate, domains.

However, not all philosophers agree with that conclusion. For example Hilary Putnam (2002) has argued that the 'fact–value' dichotomy originates from an impoverished empiricist conception of 'fact' and an equally impoverished positivist understanding of 'value'. Both ideas seem to be entangled and "a proper understanding of social and scientific change requires the abandonment of this dichotomy" (Callon 1986, but see Abi-Rached 2008). We will discuss this point further in the conclusion.

9.3.2 Problems of the "Standard Model"

The clear-cut analytic distinction between descriptive and prescriptive statements does not map one-to-one to the distinction between the steps "goal setting" and "goal verification" in the medical domain. Two problems have to be discussed here: First, both steps involve normative and empirical considerations. Second, interventions that concern capacities relevant for goal setting such as self-regulation outlined above provide further complications that undermine the significance of the clear analytical distinction between normative and empirical statements.

Table 9.1 Empirical and normative elements of goal setting and goal verification

	Normative challenge	Empirical challenge
Step 1: Goal setting	Is the goal desirable?	Is reaching the goal possible?
	Does this benefit outweigh the harm associated with reaching the goal?	
Step 2: Goal verification	Are the threshold values that determine "positive" or "negative" verification adequate?	Are the testing tools adequate to measure goal fulfillment?
	Are the verification tools ethically acceptable?	Is the goal actually met?

The first problem is of less relevance to the core question of whether neuroscientific findings have normative implications. It just points to the fact that both "goal setting" and "goal evaluation" involve empirical and normative practices – although the weights are distributed unevenly, as Table 9.1 outlines. Goal setting is primarily a normative endeavor, as the key questions to answer are normative ones: Is the goal desirable (for the patient, but also taking into account a broader perspective that involves the patient's social embedding and maybe also societal aspects such as costs)? Does this benefit outweigh risks associated with the intervention? Those questions require a "yes" in order to justify an intervention. However, also determining the therapeutic goal involves an empirical component—in particular, an assessment of whether the goal actually can be reached in the single case with sufficient probability. But the empirical component by itself does not answer the normative questions mentioned above. For instance, interventions with a low probability of success can be justified in certain cases (assuming that aspects related to distributive justice have been taken into account; Bunnik et al. 2018).

Goal evaluation is primarily an empirical endeavor. One has to use the correct tools in order to assess the success of the intervention and one indeed has to determine that success has been reached in order to claim that the intervention actually worked. However, goal verification also implies normative questions on the methodological level; e.g. when determining threshold levels for "abnormal" values the measurement device generates. In addition, tools for assessing an intervention should be ethically acceptable (e.g., by not inflicting unnecessary pain).

Usually, the intertwining of empirical and normative aspects in goal setting and goal verification can be handled on the practical level of most medical decision-making processes. However, the situation becomes more complex when the intervention concerns moral competences of the person that are directly linked to the normative challenges mentioned above. By "moral competences" we refer to a rich set of abilities that can be summarized by the model of "moral intelligence" (Tanner and Christen 2013) and that involve several of the mechanisms discussed in Sect. 9.2, in particular self-regulation and agency. All four normative challenges (see Table 9.1) can be affected by this second problem: There may be significant disagreement among the involved stakeholders (medical experts, patient, and next of kin) regarding the question, whether the intervention goal (e.g. increasing agency or self-regulation, reducing hypomania, increasing sexual drive and the like) is desirable or regarding the evaluation of risks (of e.g. becoming morally hypersensitive, hypersexual, impulsive). The assessment of risks and benefits as well as the determination of threshold values may strongly depend on societal considerations where no consensus exists. In addition, the ethical acceptability of evaluation tools may suffer from divergent evaluations from both stakeholders and societal considerations.

In particular, the following problems could emerge: First, the affected person (patient) does not think that the goal is desirable – but other persons believe that approval of desirability would be present *after* the intervention (just as in cases of certain manic states characterized by excessive involvement in pleasurable activities that have a high potential for negative consequences such as engaging in gambling, sexual indiscretions and the like (see e.g. Synofzik et al. 2012)). Then the intervention would require disregarding the capacity of the patient to make an informed decision with respect to the intended behavioral change due to his illness. Second, the affected person thinks that the goal is desirable – but the intervention leads to a situation where the person *post factum* disagrees with the desirability of the goal – for example because the person thinks that one should reach moral improvement through right reasons and not through brain interventions. In other words, the person may disagree with the means that led to the behavioral change after the intervention. This is particularly true, if she follows a deontological, virtue ethics, or other non-consequentialist understanding of morality, whereas a consequentialist may be more willing to *post factum* accept the kind of intervention used. This imposes a conflict concerning goal verification: the person will not agree with what the verification tells. Third, interaction effects may occur, in particular if the goal of the intervention is to benefit a whole system (e.g., social environment of the person, maybe even societal institutions, by improving impulse control). The problem here is that there is not a "one-way road" from intervening on the biological system all the way up to a desired social change. Changes on all levels (biological, psychological, social, societal) will feed back onto other levels. What looks like a simple feedforward process from biological intervention up to desired societal changes is actually a complex system that requires a much more expanded view of the system that is changed by an intervention (see the considerations in the next section).

9.3.3 Challenges of Neuro-modulating Moral Behavior

In Sect. 9.2, we argued that it seems conceivable that in the future, we can selectively interfere with the neural circuits to improve moral capacities possibly relating to self-regulation, sense of agency, which may also have positive effects on cooperation, trust, generosity or moral judgment and behavior in general. Albeit the time is not yet ripe for targeted interventions to improve self-regulatory abilities, we base our considerations on the assumption that neuromodulation of the "moral brain" might soon become a realistic intervention option.

9.3.3.1 Justifying Targeted Neuro-moral Interventions

Probably the most pressing consideration against the use of neuromoral interventions involves potential harms to the individual. The underlying notion suggests that there are simply no safe means such that moral changes are reliable and free from severe side effects. While the experiences made with psychotropic drugs such as ataractics make it plausible that pharmacological interventions are likely to have side effects and may fail in a considerable number of cases, it is not impossible that other means of interventions (e.g. DBS) may have more reliable effects.

Strikingly, the domain of violent behavioral manifestations is probably one of the few contexts for which a widely shared consensus on the acceptability of the intervention exists (De Ridder et al. 2009). In particular, if aggressive tendencies are at stake that are linked to direct physical harm towards third parties, as in some forms of antisocial personality disorders. Justified interventions appeal therefore just to interventions that comprise more or less uncontroversial "bad behaviors". Needless to say, the number of contexts for which one anticipates such a consensual classification is presumably rather low. Taken together, interventions with a moral aim may only be justified in cases where one addresses severe moral deficiencies with acute risks for third parties, such as attempts to correct severe aggression through DBS (Benedetti-Isaac et al. 2015).

9.3.3.2 Problems Beyond Patient Harm

However, given the numerous disastrous circumstances that are mainly man-made, the aim of improving humankind towards being more cooperative and peaceable is appealing (Persson and Savulescu 2012; Shook 2012). Therefore, it seems tempting to pursue the goals already chronicled by Skinner and many others before and after him. However, it is important to note that for many such ethically desirable behaviors, the situation is substantially more complex as many of these terms are inherently vague, depend on cultural contexts and have dramatically changed over time. Here, the "standard model" is confronted with additional problems when it comes to targeted interventions into the neural system for changing moral behavior.

The first, general point to make is that the neuropsychiatric domain poses exceptional challenges when making clear-cut distinctions between the empirical and the normative domain. This refers to psychiatry's subliminal claims about the definition of socially acceptable and desirable behaviors. No other medical discipline operates in such an area of conflict between biology, philosophy and human value judgment. Accordingly, diagnostic criteria have not only been adjusted due to cultural change but in addition, have been the object of considerable controversies. It is plausible to assume that extending the therapeutic spectrum of neuropsychiatric interventions explicitly into the moral domain will lead to much stronger controversies. The question "Is the goal desirable?" will be hard to answer as soon as the expression "improve ethical behavior" is filled with content.

Then, on the validation side, there would be a substantial difficulty of assessing moral capacities in a reliable and valid way, given the scarcity of instruments in that domain. The reason for this scarcity is that the empirical postoperative analysis is confronted with a difficult measurement problem. Complex interventions may work best if tailored to local circumstances rather than being completely standardized. But lack of standardization would confound empirically robust research designs. The rationale for a complex intervention, the changes that are expected, and how change is to be achieved may not be clear at the outset, particularly if moral capacities are part of the intervention goals.

In addition to this methodological challenge, assessing the outcome of changes to peoples' moral capacities might very well be undermined by problems of acceptability (political or ethical objections to the intervention goals), recruitment, replication, randomization, instrument scarcity and smaller than expected effect sizes (see e.g. Craig et al. 2008). Moreover, the issue of the putative reversibility of interventions such as DBS would be an important factor for the acceptability of research into this direction. Given the vagueness of the primary study outcome (moral capacities), statistical analysis would most probably have to be replaced or adjusted by a hermeneutical, philosophical and psychological evaluation. A long-term follow-up period might also be necessary to determine whether outcomes predicted by interim or surrogate measures do occur or whether short term changes persist (Craig et al. 2008).

A further caveat relates to intervention optimization: what is the optimal level of a moral capacity such as moral sensitivity (the ability to perceive moral issues when they arise in practice; Ineichen et al. 2017) between the poles of moral blindness to moral hypersensitivity? The key to be clear about is how much change or adaption is permissible and to record outcome variation. Owing to the problem of reliably measuring such complex changes as well as strict standardization, measurement fidelity is not straightforward in the context of such complex interventions.

In summary, major prerequisites for developing and evaluating complex interventions into the "moral brain" generally include a solid theoretical understanding of both the intervention goal (which is hard to achieve for moral capacities, due to their pluralistic notions) and how the intervention causes change (which is also hard to achieve, given the scarce knowledge on the neural underpinnings of moral capacities). As a consequence, lack of effectiveness may reflect implementation failure, genuine ineffectiveness or outcome measurement insensitivity. Due to higher-level processes involved in moral capacities, variability in individual outcomes is highly probable. As a consequence, sample sizes would need to be large to take into account extra variability, and mixed-methods designs should be considered. Given the vagueness of the primary outcome, multiple measures will be needed rather than a single outcome measurement.

9.4 Conclusion

Recent neurotechnological advances have shown their potential in modulating aberrant physiological processes. In future, reliable and safe neurointerventions will be present for targeted procedures aiming at altering moral abilities of individuals. Through the provision of insight in identifying important neuroanatomical nodes that achieve neural computations relevant for moral abilities of patients, deep brain stimulation has offered a first understanding of how to intervene into the complex infrastructure of socio-moral signal processing. In particular, the impact of neurointerventions on agency and self-regulation as two relevant constituents of one's personality that may influence moral behavior, has been outlined. The normative analysis revealed that the intertwining of empirical and normative aspects in goal setting and goal verification, a phenomenon that can be handled rather well in most medical decision-making processes, will face substantial problems when altering moral capacities becomes a therapeutic practice. Even if maintaining an analytical distinction between the normative and empirical perspective appears reasonable, the practical handling of neuromodulation techniques that involve normative intervention goals is likely to show the limit of this theoretical distinction. It remains an open question whether the unity of the Thomistic worldview that provided impressive coherence to the medieval era and that was later replaced by a modern trichotomy regains relevance. In particular, the distinction between the moral and empirical domain is challenged when facing the increased intertwining of empirical and normative aspects in goal setting and goal verification in the context of neuromodulating moral behavior.

We close with the realization that through recent advances in neurotechnologies, neuroscience has shown to provide a new and refined basis for understanding the neural foundations of human moral abilities. Regarding the normative implications of neuroscience it can be concluded that neuroscience does have normative implications in at least two respects: when therapeutic interventions alter moral capacities of patients and when neuro-modulatory practices impact on decision-making processes for justifying medical interventions.

References

Abi-Rached, J. M. 2008. The implications of the new brain sciences: the 'decade of the brain'is over but its effects are now becoming visible as neuropolitics and neuroethics, and in the emergence of neuroeconomies. *EMBO reports* 9 (12): 1158–1162.

Avena-Koenigsberger, A., B. Misic, and O. Sporns. 2018. Communication Dynamics in Complex Brain Networks. *Nature Reviews Neuroscience* 19 (1): 17.

Ballanger, B., T. van Eimeren, E. Moro, A.M. Lozano, C. Hamani, P. Boulinguez, G. Pellecchia, S. Houle, Y.Y. Poon, A.E. Lang, and A.P. Strafella. 2009. Stimulation of the Subthalamic Nucleus and Impulsivity: Release Your Horses. *Annals of Neurology* 66 (6): 817–824.

Bandura, A., G.V. Caprara, C. Barbaranelli, M. Gerbino, and C. Pastorelli. 2003. Role of Affective Self-Regulatory Efficacy in Diverse Spheres of Psychosocial Functioning. *Child Development* 74 (3): 769–782.

Baumeister, R.F., M. Gailliot, C.N. DeWall, and M. Oaten. 2006. Self-Regulation and Personality: How Interventions Increase Regulatory Success, and How Depletion Moderates the Effects of Traits on Behavior. *Journal of Personality* 74 (6): 1773–1802.

Benedetti-Isaac, J.C., M. Torres-Zambrano, A. Vargas-Toscano, E. Perea-Castro, G. Alcalá-Cerra, L.L. Furlanetti, T. Reithmeier, T.S. Tierney, C. Anastasopoulos, E.T. Fonoff, and W.O. Contreras Lopez. 2015. Seizure Frequency Reduction After Posteromedial Hypothalamus Deep Brain Stimulation in Drug-Resistant Epilepsy Associated with Intractable Aggressive Behavior. *Epilepsia* 56 (7): 1152–1161.

Berker, S. 2009. The Normative Insignificance of Neuroscience. *Philosophy & Public Affairs* 37 (4): 293–329.

Bratman, M.E. 2000. Reflection, Planning, and Temporally Extended Agency. *Philosophical Review*: 35–61.

Bunnik, E.M., N. Aarts, and S. van de Vathorst. 2018. Little to Lose and No Other Options: Ethical Issues in Efforts to Facilitate Expanded Access to Investigational Drugs. *Health Policy*. 2018 Jun 18. pii: S0168-8510(18)30184-2. doi: 10.1016/j.healthpol.2018.06.005. [Epub ahead of print].

Callon, M. 1986. The Sociology of an Actor-Network: The Case of the Electric Vehicle. In *Mapping the Dynamics of Science and Technology*, 19–34. London: Palgrave Macmillan.

Casebeer, W.D. 2003. Moral Cognition and Its Neural Constituents. *Nature Reviews Neuroscience* 4 (10): 840.

Casey, B.J., J.N. Epstein, J. Buhle, C. Liston, M.C. Davidson, S.T. Tonev, J. Spicer, S. Niogi, A.J. Millner, A. Reiss, A. Garrett, S.P. Hinshaw, L.L. Greenhill, K.M. Shafritz, A. Vitolo, L.A. Kotler, M.A. Jarrett, and G. Clover. 2007. Frontostriatal Connectivity and its Role in Cognitive Control in Parent-Child Dyads with ADHD. *American Journal of Psychiatry* 164 (11): 1729–1736.

Castrioto, A., E. Lhommée, E. Moro, and P. Krack. 2014. Mood and Behavioural Effects of Subthalamic Stimulation in Parkinson's Disease. *The Lancet Neurology* 13 (3): 287–305.

Cathcart, T. 2013. *The Trolley Problem, or Would You Throw the Fat Guy Off the Bridge?: A Philosophical Conundrum*. Workman Publishing.

Christen, M., and S. Müller. 2012. Current Status and Future Challenges of Deep Brain Stimulation in Switzerland. *Swiss Medical Weekly* 142: w13570.

Christen, M., and M. Regard. 2012. Der 'unmoralische Patient'. Eine Analyse der Nutzung hirnverletzter Menschen in der Moralforschung. *Nervenheilkunde* 31: 209–214.

Craig, P., P. Dieppe, S. Macintyre, S. Michie, I. Nazareth, and M. Petticrew. 2008. Developing and Evaluating Complex Interventions: The New Medical Research Council Guidance. *BMJ* 337: a1655.

Da Cunha, C., S.L. Boschen, A.A. Gómez, E.K. Ross, W.S. Gibson, H.K. Min, K.H. Lee, and C.D. Blaha. 2015. Toward Sophisticated Basal Ganglia Neuromodulation: Review on Basal Ganglia Deep Brain Stimulation. *Neuroscience & Biobehavioral Reviews* 58: 186–210.

De Ridder, D., B. Langguth, M. Plazier, and T. Menowsky. 2009. Moral Dysfunction and Potential Treatments. In *The Moral Brain*, ed. J. Verplaetse et al., 155–183. Berlin: Springer.

Deuschl, G., C. Schade-Brittinger, P. Krack, J. Volkmann, H. Schäfer, K. Bötzel, C. Daniels, A. Deutschländer, U. Dillmann, W. Eisner, and D. Gruber. 2006. A Randomized Trial of Deep-Brain Stimulation for Parkinson's Disease. *New England Journal of Medicine* 355 (9): 896–908.

Dolan, R.J. 1999. On the Neurology of Morals. *Nature Neuroscience* 2 (11): 927.

Earp, B.D., T. Douglas, and J. Savulescu. 2017. Chapter 11: Moral Neuroenhancement. In *The Routledge Handbook of Neuroethics*, ed. Johnson LSM and K.S. Rommelfanger. New York: Routledge.

El-Hai, J. 2005. *The Lobotomist: A Maverick Medical Genius and His Tragic Quest to Rid the World of Mental Illness*. Hoboken, NJ: Wiley.

Follett, K.A., F.M. Weaver, M. Stern, K. Hur, C.L. Harris, P. Luo, W.J. Marks Jr., J. Rothlind, O. Sagher, C. Moy, and R. Pahwa. 2010. Pallidal Versus Subthalamic Deep-Brain Stimulation for Parkinson's Disease. *New England Journal of Medicine* 362 (22): 2077–2091.

Foot, P. (1967). *The Problem of Abortion and the Doctrine of Double Effect*.

Frank, M.J., J. Samanta, A.A. Moustafa, and S.J. Sherman. 2007. Hold Your Horses: Impulsivity, Deep Brain Stimulation, and Medication in Parkinsonism. *Science* 318 (5854): 1309–1312.

Fumagalli, M., and A. Priori. 2012. Functional and Clinical Neuroanatomy of Morality. *Brain* 135 (7): 2006–2021.

Gehlen, A. 1940. *Der Mensch. Seine Natur und seine Stellung in der Welt*.

Gilbert, F., J.N.M. Viaña, and C. Ineichen. 2018. Deflating the "DBS causes personality changes" bubble. *Neuroethics* 1–17.

———. 2020. Deflating the Deep Brain Stimulation Causes Personality Changes Bubble: The Authors Reply. *Neuroethics* 1–12.

Glannon, W., and C. Ineichen. 2016. Philosophical Aspects of Closed-Loop Neuroscience. In *Closed Loop Neuroscience*, 259–270.

Greene, J.D., R.B. Sommerville, L.E. Nystrom, J.M. Darley, and J.D. Cohen. 2001. An fMRI Investigation of Emotional Engagement in Moral Judgment. *Science* 293 (5537): 2105–2108.

Gross, J.J. 1998. The Emerging Field of Emotion Regulation: An Integrative Review. *Review of General Psychology* 2 (3): 271.

Hack, N., U. Akbar, A. Thompson-Avila, S.M. Fayad, E.M. Hastings, E. Moro, K. Nestor, H. Ward, M. York, and M.S. Okun. 2014. Impulsive and Compulsive Behaviors in Parkinson Study Group (PSG) Centers Performing Deep Brain Stimulation Surgery. *Journal of Parkinson's Disease* 4 (4): 591–598.

Haggard, P. 2017. Sense of Agency in the Human Brain. *Nature Reviews Neuroscience* 18 (4): 196.

Haidt, J. 2007. The New Synthesis in Moral Psychology. *Science* 316 (5827): 998–1002.

Hamani, C., M.P. McAndrews, M. Cohn, M. Oh, D. Zumsteg, C.M. Shapiro, R.A. Wennberg, and A.M. Lozano. 2008. Memory Enhancement Induced by Hypothalamic/Fornix Deep Brain Stimulation. *Annals of Neurology* 63 (1): 119–123.

Hariz, M. 2012. Twenty-Five Years of Deep Brain Stimulation: Celebrations and Apprehensions. *Movement Disorders* 27 (7): 930–933.

Harris, S. 2011. *The Moral Landscape: How Science Can Determine Human Values*. Simon and Schuster.

Ineichen, C., and M. Christen. 2017. Hypo-and Hyperagentic Psychiatric States, Next-Generation Closed-Loop DBS, and the Question of Agency. *AJOB Neuroscience* 8 (2): 77–79.

Ineichen, C., H. Baumann-Vogel, and M. Christen. 2016. Deep Brain Stimulation: In Search of Reliable Instruments for Assessing Complex Personality-Related Changes. *Brain Sciences* 6 (3): 40.

Ineichen, C., M. Christen, and C. Tanner. 2017. Measuring Value Sensitivity in Medicine. *BMC Medical Ethics* 18 (1): 5.

Ineichen, C., N.R. Shepherd, and O. Sürücü. 2018. Understanding the Effects and Adverse Reactions of Deep Brain Stimulation: Is It Time for a Paradigm Shift Towards a Focus on Heterogenous Biophysical Tissue Properties Instead of Electrode Design Only? *Fontiers in Human Neuroscience*. under review.

Jahanshahi, M., I. Obeso, C. Baunez, M. Alegre, and P. Krack. 2015. Parkinson's Disease, the Subthalamic Nucleus, Inhibition, and Impulsivity. *Movement Disorders* 30 (2): 128–140.

Jonides, J., and D.E. Nee. 2006. Brain Mechanisms of Proactive Interference in Working Memory. *Neuroscience* 139 (1): 181–193.

Jonides, J., C. Marshuetz, E.E. Smith, P.A. Reuter-Lorenz, R.A. Koeppe, and A. Hartley. 2000. Age Differences in Behavior and Pet Activation Reveal Differences in Interference Resolution in Verbal Working Memory. *Journal of Cognitive Neuroscience* 12 (1): 188–196.

Kahneman, D., and D. Chajczyk. 1983. Tests of the Automaticity of Reading: Dilution of Stroop Effects by Color-Irrelevant Stimuli. *Journal of Experimental Psychology: Human Perception and Performance* 9 (4): 497.
Kennett, J. 2001. *Agency and Responsibility: A Common-Sense Moral Psychology*. Oxford University Press.
Kober, H., P. Mende-Siedlecki, E.F. Kross, J. Weber, W. Mischel, C.L. Hart, and K.N. Ochsner. 2010. Prefrontal–Striatal Pathway Underlies Cognitive Regulation of Craving. *Proceedings of the National Academy of Sciences* 107 (33): 14811–14816.
Kopell, B.H., and B.D. Greenberg. 2008. Anatomy and Physiology of the Basal Ganglia: Implications for dbs in Psychiatry. *Neuroscience & Biobehavioral Reviews* 32 (3): 408–422.
Lambert, C., L. Zrinzo, Z. Nagy, A. Lutti, M. Hariz, T. Foltynie, B. Draganski, J. Ashburner, and R. Frackowiak. 2015. Do We Need to Revise the Tripartite Subdivision Hypothesis of the Human Subthalamic Nucleus (STN)? Response to Alkemade and Forstmann. *NeuroImage* 110: 1–2.
Laxton, A.W., D.F. Tang-Wai, M.P. McAndrews, D. Zumsteg, R. Wennberg, R. Keren, et al. 2010. A Phase I Trial of Deep Brain Stimulation of Memory Circuits in Alzheimer's Disease. *Annals of Neurology* 68 (4): 521–534.
Liston, C., M.M. Miller, D.S. Goldwater, J.J. Radley, A.B. Rocher, P.R. Hof, and B.S. McEwen. 2006. Stress-Induced Alterations in Prefrontal Cortical Dendritic Morphology Predict Selective Impairments in Perceptual Attentional Set-Shifting. *Journal of Neuroscience* 26 (30): 7870–7874.
Macmillan, M. 2000. *An Odd Kind of Fame. Stories of Phineas Gage*. Cambridge, MA: MIT-Press.
Mele, A.R. 2001. *Autonomous Agents: From Self-Control to Autonomy*. Oxford University Press on Demand.
Müller, S., and M. Christen. 2011. Deep Brain Stimulation in Parkinsonian Patients—Ethical Evaluation of Cognitive, Affective, and Behavioral Sequelae. *AJOB Neuroscience* 2 (1): 3–13.
Nambu, A. 2008. Seven Problems on the Basal Ganglia. *Current Opinion in Neurobiology* 18 (6): 595–604.
Olanow, C.W., F. Stocchi, and A. Lang, eds. 2011. *Parkinson's Disease: Non-motor and Non-dopaminergic Features*. Wiley.
Persson, I., and J. Savulescu. 2012. *Unfit for the Future: The Need for Moral Enhancement. Uehiro Series in Practical Ethics*. Oxford: Oxford University Press.
Petrides, M., and D.N. Pandya. 1999. Dorsolateral Prefrontal Cortex: Comparative Cytoarchitectonic Analysis in the Human and the Macaque Brain and Corticocortical Connection Patterns. *European Journal of Neuroscience* 11 (3): 1011–1036.
Poldrack, R.A., and M.J. Farah. 2015. Progress and Challenges in Probing the Human Brain. *Nature* 526 (7573): 371.
Putnam, H. 2002. *The Collapse of the Fact/Value Dichotomy and Other Essays*. Harvard University Press.
Rüfner, V. 1964. Das Personsein im Lichte gestalthaft-genetischer Betrachtungsweise: Im Hinblick auf das religiöse Erleben. *Archiv für Religionspsychologie/Archive for the Psychology of Religion*: 231–248.
Schapira, A.H., K.R. Chaudhuri, and P. Jenner. 2017. Non-motor Features of Parkinson Disease. *sNature Reviews Neuroscience* 18 (7): 435.
Sesack, S.R., and A.A. Grace. 2010. Cortico-Basal Ganglia Reward Network: Microcircuitry. *Neuropsychopharmacology* 35 (1): 27–47.
Shook, J.R. 2012. Neuroethics and the Possible Types of Moral Enhancement. *AJOB Neuroscience* 3 (4): 3–14.
Skinner, B.F. 1972. *Beyond Freedom and Dignity*. New York: Bantam Books.
Somerville, L.H., T.F. Heatherton, and W.M. Kelley. 2006. Anterior Cingulate Cortex Responds Differentially to Expectancy Violation and Social Rejection. *Nature Neuroscience* 9 (8): 1007–1008.

Synofzik, M., T.E. Schlaepfer, and J.J. Fins. 2012. How Happy Is Too Happy? Euphoria, Neuroethics, and Deep Brain Stimulation of the Nucleus Accumbens. *AJOB Neuroscience* 3 (1): 30–36.

Tangney, J.P., R.F. Baumeister, and A.L. Boone. 2004. High Self-Control Predicts Good Adjustment, Less Pathology, Better Grades, and Interpersonal Success. *Journal of Personality* 72 (2): 271–324.

Tanner, C., and M. Christen. 2013. Moral Intelligence: A Framework for Understanding Moral Competences. In *Empirically Informed Ethics: Morality Between Facts and Norms*, ed. M. Christen, J. Fischer, M. Huppenbauer, C. Tanner, and C. van Schaik, 119–136. Berlin: Springer.

Temel, Y., A. Blokland, H.W. Steinbusch, and V. Visser-Vandewalle. 2005. The Functional Role of the Subthalamic Nucleus in Cognitive and Limbic Circuits. *Progress in Neurobiology* 76 (6): 393–413.

Thomson, J.J. 1985. The Trolley Problem. *The Yale Law Journal* 94 (6): 1395–1415.

Tremblay, L., Y. Worbe, S. Thobois, V. Sgambato-Faure, and J. Féger. 2015. Selective Dysfunction of Basal Ganglia Subterritories: From Movement to Behavioral Disorders. *Movement Disorders* 30 (9): 1155–1170.

Valenstein, E.S. 1986. *Great and Desperate Cures: The Rise and Decline of Psychosurgery and Other Radical Treatments for Mental Illness*. Basic Books.

Wallace, R.J. 1994. *Responsibility and the Moral Sentiments*. Harvard University Press.

Weaver, F.M., K. Follett, M. Stern, K. Hur, C. Harris, W.J. Marks, J. Rothlind, O. Sagher, D. Reda, C.S. Moy, and R. Pahwa. 2009. Bilateral Deep Brain Stimulation vs Best Medical Therapy for Patients With Advanced Parkinson Disease: A Randomized Controlled Trial. *JAMA* 301 (1): 63–73.

Welt, L. 2009. Über Charakterveränderungen des Menschen infolge von Läsionen des Stirnhirns. *Deutsches Archiv für klinische Medicin* 42: 339–390.

Wikström, P.O.H. 2005. The Social Origins of Pathways in Crime: Towards a Developmental Ecological Action Theory of Crime Involvement and Its Changes. *Integrated Developmental and Life-Course Theories of Offending* 14: 211–245.

Young, Liane, and James Dungan. 2012. Where in the Brain Is Morality? Everywhere and Maybe Nowhere. *Social Neuroscience* 7 (1): 1–10.

Youngerman, Brett E., Andrew K. Chan, Charles B. Mikell, Guy M. McKhann, and Sameer A. Sheth. 2016. A Decade of Emerging Indications: Deep Brain Stimulation in the United States. *Journal of Neurosurgery* 125 (2): 461–471.

Chapter 10
Autistic Moral Agency and Integrative Neuroethics

Bongrae Seok

Abstract This chapter explores three models (mutual independence, limited collaboration, and constructive integration) of interdisciplinary interaction between neuroscience and ethics and specifies three possible ways (solving ethical issues in specific contexts, developing normative standards that can be used to regulate and evaluate the behaviors of a group of individuals with particular cognitive abilities or disabilities, and identifying and correcting faulty moral intuitions) neuroscience can contribute to normative discourse of ethics. Among the three models, the author discusses and analyzes constructive integration wherein neuroscience can contribute to the development of a normative standard that refers to a group of individuals under particular psychological conditions. By surveying and analyzing recent studies of neuroscience, specifically neuroimaging studies on cognitive empathy and emotional empathy, the author argues that neuroscience can be integrated with ethics in developing a normative standard for autistic moral agency. The author also argues that, in developing and justifying a normative standard, its psychological relevance should be considered. Since a normative standard relates to a group of individuals, consideration of their cognitive and emotional abilities is critically important. In this regard, integration of neuroscience and ethics can be understood as the theoretical effort to bring neuroscience to the discussion of normative rules and standards that can be practiced by a particular group of individuals.

Keywords Moral agency · Autism · Emotional empathy · Cognitive empathy · Theory of mind · Neuroethics

B. Seok (✉)
Associate Professor of Philosophy, Alvernia University, Reading, PA, USA
e-mail: bongrae.seok@alvernia.edu

G. S. Holtzman, E. Hildt (eds.), *Does Neuroscience Have Normative Implications?*, The International Library of Ethics, Law and Technology 22,
https://doi.org/10.1007/978-3-030-56134-5_10

10.1 Introduction

One of the most intensely debated topics in philosophy today is the relevance of natural science to philosophy in the construction of normative standards of ethics. Can empirical studies of the mind and the brain be used to develop theories of right and wrong? Specifically, do cognitive psychology and neuroscience contribute to normative or axiological theories in philosophy? In this chapter, I will discuss whether and how neuroscience contributes to normative ethics. I will survey and explain different forms of the interactive or integrative relation between empirical science and philosophy and discuss the particular contribution neuroscience makes to normative ethics, specifically in the context of moral agency (the standard of being a morally capable and responsible actor or decision maker). Often, empirical studies of natural science provide counterexamples that can be used to discard conceptually possible but physically or psychologically unrealistic theories of philosophy. This type of empirical evaluation or falsification is particularly vital in normative ethics because ethical principles provide prescriptive rules or norms that one should follow, observe, and respect. Any moral principle or rule that requires physically or psychologically unrealistic abilities cannot be accepted as a practical norm of human conduct since it is irrational to ask people to follow moral rules that they have no full psychological ability to follow. Neuroscience can help us analyze and dismiss normative principles that assume and propose improbable or unrealistic moral abilities of the human mind.

In addition to the falsifying and screening roles, neuroscience can play constructive roles in normative ethics. It can help ethicists to develop a normative standard of moral agency for a particular group of individuals (i.e., what these individuals should do as morally capable agents and what their moral responsibility amounts to). In this chapter, I will discuss how a normative standard for a group of individuals can be constructed by following empirical studies of neuroscience, specifically how a standard of autistic moral agency (i.e., what autistic individuals should do morally and how they are morally responsible for their actions and decisions) can be developed. Recent studies of neuroscience provide valuable information about the psychological details of autism and autistic individuals' cognitive impairment, yet they also demonstrate autistic individuals' distinctive moral ability, which intuitive theories and conceptual analyses of moral philosophy do not clearly and fully identify. Neuroscience has the empirical advantage to analyze and construct the standard of moral praiseworthiness and blameworthiness of actions and decisions made by autistic individuals. That is, neuroscience contributes to normative ethics in guiding and developing a standard that is tailored to the moral ability and responsibility of autistic individuals. I will argue that neuroscience is not only empirically relevant but also normatively significant in our understanding of autistic moral agency.

To understand the normative significance of empirical studies, a brief survey of some philosophical background will be helpful. The relevance of empirical science in rational justification of normative standards (ideal values or prescriptive rules that regulate certain behaviors or decisions) is one of the most frequently discussed topics in Anglo-American philosophy. When Quine (1969) proposed "naturalized epistemology" (the thesis that scientific [i.e., descriptive] and philosophical [i.e., prescriptive] studies of knowledge can be integrated), he argued for the importance of empirical studies, such as psychological studies of cognition, in the rational and normative justification of knowledge. Dretske (1981), Goldman (1989, 1992, 2012), and Kornblith (1985, 2007, 2014), for example, are epistemologists who believe in the significance of empirical studies in philosophical analysis of prescriptive and rational standard of knowledge.

Regarding the naturalization of ethics, i.e., empirical integration of ethics, philosophers such as Stich (1983, 1990, 2006; Sripada and Stitch 2006), and Flanagan (1991, 2017; Flanagan et al. 2014) emphasize the theoretical relevance of empirical science in philosophical discourse and promote integration and cooperation among psychology, neuroscience, and philosophy. More recently or perhaps more specifically, Appiah (2010), Churchland (2011), Harris (2010), Kahane (2013), Knobe (2003a, b, 2006), Mikhail (2011), Nichols (2004), and Prinz (2004, 2007) actively develop sustainable and meaningful interdisciplinary links between empirical sciences (psychology and neuroscience) and moral philosophy.[1] It is not just philosophers who are interested in interdisciplinary study of morality. Psychologists such as Greene (2008, 2014), Haidt (2001, 2012; Haidt and Joseph 2007), and Hauser (2006) integrate empirical science of the mind/brain and philosophical discussion of morality in their studies of moral judgment and moral emotion. But what is the ideal relation between ethics and empirical sciences, and how does the latter contribute to the former?

In the first part of the chapter, I will explore interdisciplinary relations between neuroscience and ethics and discuss three possible options: mutual independence, limited collaboration, and constructive integration. Among these options, I will discuss and analyze the third option of interdisciplinary cooperation between neuroscience and ethics. I will argue that neuroscience can contribute to constructive moral theorizing, particularly in the development of normative standards regarding moral agency. I will use brain imaging studies of empathy as an example and explain that brain imaging data can be utilized in developing theories of autistic moral agency, i.e., theories that discuss whether and how autistic individuals are morally capable agents. If we characterize moral agency as an ability that includes a clear understanding of and appropriate reaction to others' behaviors and their inner cognitive and emotional states, making moral decisions and developing moral judgments are particularly challenging to those individuals whose theory of mind ability is limited. Autistic individuals, according to many psychological studies (Baron-Cohen 1995, 2011; Baron-Cohen et al. 1985; Senju et al. 2009), have great difficulty in understanding and reacting appropriately to others' actions by considering their motivational, intentional, and emotional states. These socio-cognitive difficulties, however, do not imply that autistic individuals are immoral or amoral.

[1] See Gert (2012), Kamm (2009), Levy (2009), and Schirmann (2013) for general reviews.

Regarding the moral abilities of autistic individuals, Kennett (2002) argues, on the basis of her philosophical analysis and psychological observation, that a rule-based model of Kantianism can explain the moral behaviors of autistic individuals. According to rule-based Kantianism, individuals are morally capable if they have the ability to recognize general rules of human conduct and to act on them consistently. Kennett believes that the rule-based Kantian model provides a good explanation of the moral abilities of autistic individuals because autistic individuals can recognize general rules and follow them, but they have difficulty in understanding others' moral beliefs and intentions. Considering the cognitive orientations of autism (intact or overly active understanding of general patterns and rules but impaired abilities in social cognition, specifically in theory of mind), her Kantian approach seems reasonable and convincing. Autistic individuals tend to understand and judge others' moral or immoral actions based on general rules without relying on their understanding of others' inner intentions or beliefs. She discusses how high-functioning autistic individuals, such as Temple Grandin (1996) and Jim Sinclair (1992), are morally responsive to others and make morally sensible decisions to show that some autistic individuals are Kantian moral agents.

Kennett's Kantian approach, however, does not take full consideration of the moral abilities of autistic individuals that depend neither on theory of mind ability nor on rule-based understanding of morality. Many psychologists (Bacon et al. 1998; Blair 1996; Leslie et al. 2006; Yirmiya et al. 1992) observe that autistic individuals are able and tend to initiate helping and caring behavior independently of their lack of theory of mind abilities (i.e., abilities to understand others' behaviors by identifying and interpreting their inner psychological states) when they observe others' actual or potential pain and suffering. Specifically, recent research in social neuroscience demonstrates the existence of emotional processes such as empathic concern that functions interactively but independently of the theory of mind abilities of moral agents. By analyzing and interpreting recent studies of neuroscience, I will argue that neuroscience can contribute to the development of normative standards of autistic moral agency and that a model of moral agency based on emotional empathy is possible for autistic moral agents independently of or in addition to a Kantian model of moral agency.

10.2 Three Models of Interdisciplinary Interaction

To understand and specify how empirical studies of neuroscience relate to normative studies of ethics, it is important to consider three different options of the interdisciplinary relation.

1. Mutual Independence Model: According to the first view, ethics can maintain its theoretical autonomy in its interface with neuroscience. As many philosophers recognize, there is a major conceptual gap between value (generally defined as ideal goals and objectives of human life) and fact (generally defined as descriptive

conditions of the world including human psychology). For this reason, fact (what is happening) does not justify value (what should ideally happen), and value does not directly derive from fact. As a conceptual or normative study of values and principles, ethics is different from neuroscience, i.e., empirical studies of the brain. Since moral judgments are made on the basis of normative values, they do not need to interact with or be integrated into empirical facts. Neuroscience and ethics, therefore, maintain their autonomies through their distinctive contributions to our understanding of morality. Simply, neuroscience studies processes and functions of the brain that underlie one's understanding of moral values and one's decisions to act morally, but ethics studies what kind of values one should pursue and what kind of actions one should take. This autonomy model supports the unique normative nature of moral discourse, but this type of autonomy or independence is not what one can take or appreciate in one's effort to bring science and ethics together in their meaningful interaction to advance our understanding of morality.

2. Limited Collaboration Model: For the constructive relation between ethics and neuroscience, one can think of a model that allows empirical contribution or feedback to moral theories. One can argue that empirical theories cannot be used to confirm or support normative moral theories because empirical facts (regarding what is happening in the world) do not *directly* support normative principles (regarding what should ideally happen). The former, however, can still be used to evaluate or reject the latter. Fully confirmed empirical facts can be used to reject moral theories that conflict with empirical facts such as psychological and neurobiological facts about moral competency and agency. That is, empirical facts, on this view, should not be used to confirm or justify moral theories, but they can be used to reject moral theories. This type of limited collaboration between neuroscience and ethics can be understood analogically by considering how a scientific hypothesis is confirmed or rejected by empirical evidence. According to Popper's (1963) falsifiability thesis, empirical evidence cannot fully confirm a given hypothesis but can be used to reject it. Perhaps this limited unilateral relation between empirical evidence and confirmation of scientific theories is applicable to empirical evidence of neuroscience and confirmation of moral theories. As empirical evidence cannot fully confirm but instead can only falsify a hypothesis, neuroscience cannot fully support, but can at least reject, a moral theory as psychologically implausible or impossible. The underlying rationale for this limited interaction between neuroscience and ethics lies in a theoretical principle in ethics: "ought implies can."[2] If any moral theory or principle justifies a moral judgment that is beyond human psychological or cognitive ability, the moral theory or principle cannot be accepted. That is, one does not

[2] Usually, a ceteris paribus (other things being equal or lacking other overriding variables) condition is added to this principle. That is, this principle should apply to moral actions and duties "under natural condition." There are, however, some issues on the interpretations and exceptions of the principle. See Stern (2004) for a general discussion of "ought implies can" in ethics.

have a moral duty to follow a moral principle that asks one to engage in psychologically implausible or cognitively impossible actions.[3]

On this view of interdisciplinary interaction, the distinction between empirical fact and normative value is still maintained but the former can be used to reject the latter. Neuroscience is minimally "relevant" in constructing moral theories (particularly in developing plausible models of moral agency and moral behavior), but critically important in rejecting implausible moral theories. That is, the limited collaboration model supports somewhat restricted interaction between neuroscience and ethics. With its careful observations and advanced brain imaging technologies, neuroscience can provide empirical information about our moral abilities and make a good interdisciplinary partner for ethics. The main portion of moral theorizing, however, remains non-empirical. It comes out of conceptual analysis and axiological consideration of normative (non-empirical, non-descriptive) moral values and principles that are developed independently of empirical studies of the mind and the brain. As a result, neuroscience is somewhat, but not fully, integrated into ethics.

3. Constructive Integration Model: The third model of interdisciplinary interaction between neuroscience and ethics is more constructive and integrative than the previous two models. Empirical observations of neuroscience can be used actively in the process of building normative standards and axiological principles. That is, in this model, neurological facts can play constitutive roles in building a normative standard of human action. For example, if a neuroscientist discovers repeatedly confirmed biological conditions of moral abilities, such as the quick and effective deployment of brain activities in response to violations of reciprocity or fairness, can she use them to propose a universal moral principle of fairness that all human being should follow? The success of this type of integration lies in the fully constructive or constitutive use of empirical data of neuroscience in normative justification of ethical theories. Can neuroscience play constitutive roles in ethics in this normatively constructive way?

Kant, who distinguishes descriptive and normative conditions of justification, would be very skeptical about this possibility. Hume, who carefully separates descriptive statements from normative statements (i.e., matters of "is" from matters of "ought"), would be equally skeptical about this type of deeply constructive and fully integrative cooperation between neuroscience and ethics. As I briefly discussed above, however, many philosophers such as Dretske (1981), Goldman (1989, 1992, 2012), Kornblith (1985, 2002, 2007, 2014) and Millikan (1984), develop constructive models of interaction between science (psychology) and philosophy (epistemology) in their effort to build a new form of epistemology, i.e., naturalized epistemology. According to Quine (1969), empirical science can play an important role in epistemology by developing psychologically plausible and fully practicable

[3] The principle is comparable to Flanagan's (1991, p. 32) principle of minimal psychological realism: "Make sure when constructing a moral theory that the character, decision processing, and behavior prescribed are possible, or are perceived to be possible, for creatures like us."

theories of epistemology.[4] Normative standards of knowledge, according to naturalized epistemologists, are not simply conceptual possibilities or ideal norms of human intellect that are discussed independently of the cognitive conditions and constraints of the human mind. Rather, they are the standards we use to build theories and to conduct scientific researches in our concrete cognitive and intellectual environments.[5] Suppose coherence as an epistemological standard. To justify one's knowledge, one needs to make sure that one's beliefs cohere with one another. Suppose there are 101 beliefs in one's belief system. To check the coherence of this system, how long does it take? If it takes one second to check the coherence of one belief with another belief, it will take 100! (100 factorial = 100 × 99 × 98 × 97 … × 3 × 2 × 1 = 933,262…000000: this is a number with 158 digits) seconds to check the full coherence of this belief system. That is, to check the coherence of this system, it will take longer than the life of a human being. Is coherence a norm of epistemology that can apply to and be practiced by human beings? If one considers epistemological norms in their practical conditions of use, application, and development in psychological and social contexts, information provided by empirical sciences regarding the cognitive abilities of epistemological agents is critically important in developing a feasible and practical theory of epistemology. Science should be consulted and integrated in our philosophical effort to build conceptually consistent, cognitively plausible, and fully practical theories of epistemology. Can we expect the same kind of constructive interaction between neuroscience and ethics? If naturalized epistemology via psychology is possible, is naturalized ethics via neuroscience possible too?

In the following sections, I will argue that neuroscience can contribute to constructive moral theorizing, particularly in the development of normative standards regarding moral agency. I will use brain imaging studies of empathy as an example to argue that empirical data of neuroscience can be used to develop a normative standard of autistic moral agency.

[4] Quine (1969, pp. 82–83) explains his naturalization project in the following way: "Epistemology, or something like it, simply falls into place as a chapter of psychology and hence of natural science. It studies a natural phenomenon, viz., a physical human subject. This human subject is accorded a certain experimentally controlled input – certain patterns of irradiation in assorted frequencies, for instance – and in the fullness of time the subject delivers as output a description of the three-dimensional external world and its history. The relation between the meager input and the torrential output is a relation that we are prompted to study for somewhat the same reasons that always prompted epistemology: namely, in order to see how evidence relates to theory, and in what ways one's theory of nature transcends any available evidence… But a conspicuous difference between old epistemology and the epistemological enterprise in this new psychological setting is that we can now make free use of empirical psychology."

[5] Naturalized epistemology can take many different forms. Some are more metaphysical and others are methodological. Generally naturalized epistemologists believe that knowledge or its justification is explained by such naturalistic properties as causation (Goldman 1967), reliability (Armstrong 1968; Goldman 1979; Kornblith 2002), natural functions (Millikan 1984), and information flow (Dretske 1981).

10.3 Autistic Moral Agency

As Kanner (1943) explains in his early observation of autistic behavior, children with autism are often characterized as socially and emotionally remote and self-enclosed individuals. Due to their cognitive and social tendencies of self-absorption and enclosure, people with autism are often regarded as morally passive, irrelevant, or incapable individuals. This general characterization of autism has been challenged recently. Many psychologists report that autistic individuals are capable moral agents with intact understanding of moral rules and violations. First, autistic children can distinguish moral violations from conventional violations and react to the former differently and appropriately (Blair 1996). Second, their emotional reactions and comforting behavior towards the victims of moral violations are comparable to their typically developing peers (Bacon et al. 1998; Yirmiya et al. 1992). Third, many high-functioning autistic individuals or individuals with Asperger syndrome, such as Grandin (1996) and Sinclair (1992), make quite a remarkable case for autistic moral agency with their strong sense of moral duty and responsibility. These reports and observations demonstrate that, even with the apparent lack of other regarding emotions, people with autism are morally motivated and considerate, if not fully capable, agents.

With the positive observations of autistic moral abilities, several philosophers have developed models of autistic moral agency. Kennett (2002) argues that some autistic individuals if they are morally able, are Kantian agents who are less dependent on other regarding emotions than on the sense of duty to universal moral rules. McGeer (2008) also acknowledges that autistic individuals who are morally active have a general sense of moral duty, but their sense of duty is formed and supported by their passion for the ultimate order of the universe. As Simon Baron-Cohen (2003, 2009) observes, autism is associated with systems thinking. Autistic individuals are good at finding systematic rules or organizing patterns. They are "systematizers": they think in terms of rules and patterns that govern a system, a structure, or an organization that functions under formalized processes.

The Kantian model can explain the systematizing or universalizing tendencies of autistic individuals in their moral actions and judgments. However, the model does not provide a good theoretical framework to explain the moral abilities of autistic individuals who are not highly intelligent or highly functioning. Since autism comes with different degrees and intensities and with different intellectual and communication abilities, applying the Kantian model to all autistic individuals seems problematic. In fact, most examples used to support the Kantian model of autistic moral agency come from autistic individuals who are highly intellectual or high functioning (Baron-Cohen et al. 2003; Jaarsma 2013). This subpopulation of autistic individuals is often diagnosed with Asperger's syndrome (a form of high-functioning autism without intellectual disability).[6]

[6] For example, Baron-Cohen (2009, p.72) states that "A second piece of evidence [to support his theory of autism as a tendency of systems thinking] comes from studies using the Systemizing

In contrast to this intellectual, systematizing, and universalizing approach, an affective-motivational approach to autistic moral agency is possible. McGeer (2008) explains the emotional or sentimentalist approach to moral agency through Hume's moral psychology (i.e., emotions are the primary foundation of moral disposition and moral knowledge) and contrasts it with Kant's rational moral psychology. She states that "Hume argued that the capacity to feel with and like another – to enter sympathetically into their cares and concerns – was critical for developing and maintaining an other-regarding moral agency. Kant, by contrast, was deeply disdainful of the moral importance of empathy and/or sympathy, favoring a moral psychology motivated in its purest form by a rational concern for doing one's duty" (McGeer 2008, p. 228). If an emotional approach to moral agency can be developed as a distinct philosophical theory of moral agency, is it psychologically plausible, specifically in the context of autistic moral agency?

To understand the moral abilities of autistic individuals, one needs to distinguish different forms of empathetic abilities, such as *cognitive empathy* (the ability of perspective-taking, the ability to understand others' inner mental states) and *emotional empathy* (the ability to feel others' affective states, and react to them with prosocial motivation and concern), and focus on emotional empathy as the psychological foundation of autistic moral agency.[7] Generally, autistic individuals do not have good understanding of "what" others think and feel, "how" others' inner experience feels like, and how it affects their behavior, but they have basic abilities to "sense" and to "react" to moral violations and wrongful sufferings. This basic, affective, spontaneous, and motivational reaction to others' pain and suffering, I believe, is an essential feature of autistic moral agency. Many neuroimaging studies provide empirical evidence to support the existence of this type of affective and reactive empathy and its prosocial (helping) and moral behaviors. These empirical studies play critical roles in constructing the emotional model of autistic moral agency.

Quotient (SQ). The higher your score, the stronger your drive to systemize. People with high-functioning autism or Asperger syndrome score higher on the SQ than people in the general population (Baron-Cohen et al. 2003). The above tests of systemizing are designed for children or adults with Asperger syndrome, not classic autism."

[7] Batson (2009), for example, lists eight different meanings or designations of empathy and Cuff and his colleagues (Cuff et al. 2016) report 43 different definitions of empathy. In general, however, there are three different forms of empathy (Decety and Cowell 2014, 2015; Zaki and Ochsner 2012) in psychology and neuroscience: perspective taking (cognitive empathy), experience sharing (emotional contagion), and empathic concern (other-regarding concern). In this chapter, I compare and contrast cognitive empathy (perspective taking) and emotional empathy (emotional contagion and empathic concern) to analyze autistic moral agency.

10.4 Neuroscience of Empathy

When one sees or hears others' pain, one feels, to a certain degree, their pain as if it is one's own. It seems that others' pain is felt directly as one's own pain in this highly emotional and vicarious experience of pain. According to Singer and her colleagues' study of pain perception (Singer et al. 2004; Bernhardt and Singer 2012; Engen and Singer 2013; Kanske et al. 2017), some brain regions in the pain matrix (brain regions that serve pain perception: the anterior insula, somatosensory cortices, supplementary motor area, anterior cingulate cortex, periaqueductal gray, thalamus, and cerebellum) serve both one's first-hand experience of one's own pain and one's observation of others' pain.[8] From the perspective of neural activation in these areas of the pain matrix, our perception of others' pain is empathic: one perceives others' pain by experiencing it vicariously in one's own mind. Singer and her colleagues' brain imaging studies report that the anterior insula and the anterior midcingulate cortex are particularly active in this process of empathic pain perception.

The vicarious experience of pain is a unique subtype of empathy, i.e., emotional empathy (emotional contagion and affective arousal typically observed in one's perception of others' affective states such as pain and suffering) that needs to be distinguished from cognitive empathy (one's ability to understand and interpret others' inner states, sometimes referred to as perspective-taking ability or theory of mind ability) (Decety and Cowell 2014; Decety et al. 2012; O'Brien et al. 2013).[9] Although cognitive empathy and emotional empathy are interrelated psychological processes, they are distinct forms of empathy served by different areas of the brain. Cognitive empathy is typically served by the right temporal parietal junction, the superior temporal sulcus, the anterior and posterior midline regions (Dodell-Feder et al. 2011; Schurz et al. 2014) and the ventromedial prefrontal cortex (Shamay-Tsoory et al. 2009). Emotional empathy in pain perception is served by the anterior insula, the anterior midcingulate cortex (Singer et al. 2004), and the inferior frontal gyrus (Shamay-Tsoory et al. 2009). The empathic processes of pain perception provide a good chance to study the emotional empathy and its moral psychological

---

[8] See Decety (2011) for a general review of empathic pain perception.

[9] More extensively, Decety distinguishes three different (cognitive, emotional, and motivational) aspects of empathy. For example, Decety and Cowell (2014, p. 337) state that "…empathy is a construct comprising of several dissociable neurocognitive components (emotional, motivational, and cognitive), interacting and operating in parallel fashion. The emotional component of empathy involves ability to share or become affectively aroused by others' emotions (at least in valence, tone, and relative intensity). It is commonly referred to as emotion contagion, or affective resonance, and its independent of mindreading and perspective-taking capacities. The motivational component of empathy (empathic concern) corresponds to the urge to care for another's welfare. Finally, cognitive empathy is similar to the construct of affective perspective-taking. Each of these emotional, motivational, and cognitive facets of empathy can influence moral behavior in dramatically different ways." My discussion of emotional empathy and its moral significance, in this chapter, covers broad characteristics of emotional empathy that includes Decety and Cowell's emotional and motivational dimensions of empathy.

characteristics because empathy caused by one's perception of others' emotional states associated with pain, suffering, and sadness, typically motivates one's prosocial and moral (i.e., helping and caring) behaviors (Balconi and Canavesio 2013; Eisenberg and Miller 1987; Murakami et al. 2014; Sze et al. 2012). It is important to note that emotional empathy is often regarded as a foundation of moral sense and disposition (Hoffman 2000; Hume 1739/1896). In his discussion of virtue, ancient Chinese philosopher Mencius (Mencius 2006) points out that the sense and emotion (affective arousal and other-concerning emotion) one feels when one observes another's pain is an important, perhaps essential, moral psychological foundation of Confucian virtue. He specifically emphasizes the moral significance of empathic nociception (emotional empathy in one's observation of others' pain) in one's cultivation of virtue.[10]

Psychological natures of cognitive and emotional empathy have been discussed in many studies of psychology and neuroscience. First, different processes and networks of empathy have been analyzed in many brain imaging studies (Decety and Lamm 2006; Decety et al. 2015; Decety and Jackson 2004; Decety and Meyer 2008; Lamm et al. 2007; Marcoux et al. 2014). For example, it is observed that cognitive and emotional empathy have distinct activation patterns in different areas of the brain (Nummenmaa et al. 2008) and, sometimes, support strong dissociation patterns. Shamay-Tsoory and her colleagues (Shamay-Tsoory 2011; Shamay-Tsoory et al. 2009) report a case of double dissociation (mutually exclusive patterns of activation) between the two forms of empathy in two distinct brain areas. The similar pattern of dissociation is observed (i.e., cognitive empathy and emotional empathy are differentially affected) in many psychological conditions such as schizophrenia, borderline personality disorder, schizotypy, Huntington's disease, alcoholism, and conduct disorder (Harari et al. 2010; Henry et al. 2008; Maurage et al. 2011, 2016; Montag et al. 2007; Poustka et al. 2010; Schwenck et al. 2012). For example, when Poustka and her colleagues (2010) compared cognitive and emotional empathy between two groups of adolescents, namely, individuals with autism and individuals with conduct disorder (a developmental disorder with persistent anti-social behaviors), they discovered a pattern of double dissociation between cognitive and emotional empathy. They report that the two "groups differed significantly on both components of empathy. Adolescents with ASD [autism spectrum disorder] showed impairments in cognitive empathy, but did not differ from healthy controls in emotional empathy. Adolescents with CD [conduct disorder] showed an inverted pattern of dissociation of empathy components, compared to adolescents with ASD." (Poustka et al. 2010, p. S81).

---

[10] Hume's (1739/1896) and Hoffman's (2000) discussions include a broad spectrum of empathy, including both cognitive and emotional aspects of empathy. Mencius also discusses empathy in this broad sense, but he often focuses on particular aspects of empathy in the passage (Mencius, 2A6) where he uses the example of a person observing a child coming close to a well (a dangerous place) to explain a particular form of empathy as a foundation of the Confucian virtue of benevolence (*ren*). In this passage, he emphasizes the moral significance of empathic nociception (emotional empathy in one's observation of others pain).

Second, the moral psychological natures of the two types of empathy differ as well. One of the effective ways to study the moral psychological characteristics of empathy is to observe its impairments in moral cognition and behavior. Often empathic abilities of psychopathic and autistic individuals are compared and contrasted to analyze differential contributions of cognitive and emotional empathy to moral emotions and dispositions. Some studies (Marsh et al. 2013; Seara-Cardoso et al. 2015) report that psychopaths have intact cognitive empathy but impaired emotional empathy. When individuals with psychopathic tendencies perceive, for instance, others' pain, they do not seem to be aroused by it or be motivated to care for the victims. Other studies (Dziobeck et al. 2008; Fan et al. 2014; Hadjikhani et al. 2014; Jones et al. 2010; Rogers et al. 2007; Rueda et al. 2015; Schwenck et al. 2012) report that autistic individuals have impaired cognitive empathy but intact emotional empathy. They seem to be affected by others' emotional states and sometimes react to them with careful attention. Although cognitive and emotional empathy always interact with other cognitive and emotional processes (Betti and Aglioti 2016), the general dissociation patterns suggest that cognitive empathy and emotional empathy serve different psychological functions and are instantiated in different brain areas. Most importantly, they serve different moral psychological functions in psychopathy and autism: more cognitive but less affective and motivational empathy in morally compromised behaviors of psychopathic individuals and more reactive and motivational but less cognitive empathy in morally intact behaviors of autistic individuals.

The psychological and neural dissociation between cognitive and emotional empathy suggests that emotional empathy (emotional contagion and arousal caused by one's observation of others' affective states) is a necessary component of moral agency or moral motivation. Lacking emotional empathy can result in anti-social behaviors including psychopathy and conduct disorder (Blair 2005; Schwenck et al. 2012; Segal et al. 2017). The dissociation pattern also suggests that emotional empathy is an important psychological foundation of the autistic moral agency. If autistic individuals can sense and feel others' suffering and act for others' wellbeing without cognitive empathy (Blair 1996; Leslie et al. 2006), their intact emotional empathy can explain their moral abilities fully or at least partially. Many psychologists observe that emotional empathy is strongly correlated with prosocial (helping and caring) behaviors (Balconi and Canavesio 2013; Eisenberg and Miller 1987; Murakami et al. 2014; Sze et al. 2012). That is, autistic individuals' moral abilities can be explained by the moral (other-regarding or other-caring) tendencies of emotional empathy.

Considering the differential roles of cognitive and emotional empathy in moral cognition and disposition, one could argue that autistic individuals are moral agents because their emotional empathy is intact. Because of their impaired cognitive empathy, however, their moral judgments can be affected by their limited ability to consider the full scope of intention and other inner motivations: they have difficulty distinguishing intentional violations from unintentional harm (Moran et al. 2011; Zalla et al. 2011). Nevertheless, they can clearly understand and distinguish moral transgressions (kicking someone or pulling someone's hair) from conventional transgressions (violation of general social norms such as wearing pajamas in public places, playing with food) (Blair 1996; Leslie et al. 2006). For example, in a study by Leslie and his colleagues (2006), autistic children could distinguish morally relevant pain (pain of a victim induced by a wrong behavior) from morally irrelevant pain (morally neutral or unrelated crying behavior) and react to morally relevant pain appropriately with prosocial (helping and comforting) behaviors.[11] Leslie and his colleagues' results suggest that the moral ability of autistic children does not depend on cognitive empathy or theory of mind ability. Nor does it depend on Kantian moral deliberation, considering their ages (three- to five-year-old children) and their underdeveloped ability of moral deliberation. Then, they sense and react to others' pain with empathic concern guided by emotional empathy independently of cognitive empathy. Probably, the best explanation of autistic children's perception of others' pain and the following prosocial or caring behaviors is to assume a basic form of moral sense served by emotional empathy. Autistic moral agency, therefore, is based on this limited but effective form of empathy.

10.5 Integration of Neuroscience and Ethics

With the advancement of brain imaging technologies, psychologists can distinguish different types of empathy and their moral psychological implications. This type of empirical observation also helps philosophers to develop a plausible model of autistic moral agency. Autistic individuals, unlike psychopaths, can act morally. They are capable moral agents, but their social cognitive functions are impaired. For this reason, any theory of autistic moral agency that presupposes full social cognitive ability such as theory of mind ability cannot justify its normative standard of autistic moral ability.

One can, therefore, understand that neuroscience plays the role of filtering out or rejecting implausible models of autistic moral agency that presuppose fully developed social cognitive abilities such as theory of mind ability or cognitive empathy in the minds of autistic moral agents. This interdisciplinary cooperation between neuroscience and ethics follows the limited collaboration model that I discussed in the second section of this chapter. Interaction between neuroscience and moral philosophy, however, does not stop at this limited collaboration.

Beyond this limited collaboration, there are three possible ways the integration of neuroscience and ethics can take place. First, neuroscience and ethics can be

[11] Leslie and his colleagues (Leslie et al. 2006, p. 279) state that "Both normally developing and autistic children responded more positively [i.e., approvingly] to cry baby stories indicating that their judgments distinguish between the distress of a "cry baby" and the distress of a victim. Although both the moral and the cry baby stories featured a character who starts to cry following the actions of another person, only in the moral stories can that action remotely be deemed culpable rather than a mere cause. This, in turn, suggests that the reaction to distress cues in moral transgressions is not simply of the "knee jerk" type but involves moral reasoning."

integrated on ethical issues or topics that are raised in particular contexts and situations. One of the most frequent criticisms of the integration of neuroscience and ethics (neuroscience becoming the normative science of human conduct) is neuroscience cannot discuss and provide broad and general norms that everyone should follow in most situations. In his article "What the science of morality doesn't say about morality", Abend (2013) points out that neuroscience may provide detailed arguments about a particular kind of individual moral judgments, but it does not provide general views of morality that encompass a broad range of moral phenomena. He argues that neuroscience is good at analyzing moral judgments under particular conditions, but empirical analyses of individual moral judgments do not fully justify general theories of ethics. A similar point is made by Gert (2012). As he reviews different (Appiah's 2010; Churchland's 2011; Harris's 2010) approaches neuroscience can take to the normative discourse of ethics, he emphasizes that there are diversely different viewpoints on the normative contributions of neuroscience to ethics. He then states that "The fact that those who relate neuroscience to morality differ so radically in their accounts of morality suggests that neuroscience has nothing to add to our understanding of morality as a code of conduct that everyone should follow. However, neuroscience may help explain why some people behave as they do in situations that call for moral decisions or judgments" (Gert 2012, p. 28). Although both Abend (2013) and Gert (2012) take a critical stance toward the integration of neuroscience and ethics, their criticisms seem to suggest the possibility that the integration can be successful if ethical issues or principles are discussed in particular contexts.

Because of its interest in controlled observation and careful analysis of individual moral judgments, neuroscience can discuss particular issues or conditions of ethics better than general theories of ethics such as utilitarianism and Kantian deontology. For example, in their article "What can neuroscience contribute to the debate over nudging?," Felsen and Reiner (2015) provide a good example of how neuroscience can develop normative views on ethical issues in a very specific context. They discuss the ethics of nudging (covert external influences) in a context of decision-making processes. They use empirical data from neuroscience and develop a model to show how nudges can promote or restrain autonomy. Then they explore and construct normative standards of nudging. In this concrete and specific context, neuroscience can bring integrated empirico-normative viewpoints that can be used to solve or to provide possible solutions to the ethical issues of nudging. That is, the integration of neuroscience and ethics can be promising if it discusses particular ethical issues under specific contexts or situations.

Second, the integration can have a better chance if neuroscience and ethics discuss the moral standards of a particular group of individuals under their specific psychological conditions (such as cognitive impairments, psychopathologies, and intoxications). In this chapter, I pursue this type of integration where neuroscience plays an important role in developing a moral standard suitable for autistic individuals. By providing information about the cognitive and emotional abilities of a group of individuals, neuroscience can contribute to the development of moral standards for autistic agency. If ethical norms are not applicable to people (i.e., particular individuals with their specific psychological abilities and disabilities) and their actions, they cannot function as moral standards, i.e., standards that provide practical guidance. In this regard, Aaltola's (2014), Krahn and Fenton's (2009), and Stout's (2016a, b) works are good examples of the integration where empirical studies of the mind and the brain are actively consulted to analyze and construct normative standards of autistic moral agency. Specifically, they pursue theoretical options and views of moral agency that are consistent with the emotional and empathic abilities of autistic individuals.

Third, theoretical integration has moral significance when our moral intuitions are faulty or inconsistent. One of the important contributions of neuroscience (or cognitive science, in general) to ethical debates is its critical stance toward unjustified or unfounded intuitions in moral reasoning. Some of these faulty intuitions are generated by strong emotional reactions (such as disgust and anger) and blind rejections (such as quick and spontaneous feelings of hatred) to certain behaviors or conducts. Other intuitions are less emotional but subject to equally powerful cognitive biases (such as the framing effect and the overconfidence effect).

Neuroscience often provides information on how these intuitions give rise to strong moral judgments, not to justify them but to carefully explain and analyze them. Perhaps it can also find ways to neutralize or eliminate blind, strong, and disturbing intuitions. As Levin (2011) points out, faulty and unjustified intuitions are a major issue in moral philosophy, and one of the effective ways to identify and disable them is to integrate neuroscience in moral reasoning. Since some intuitions are not transparently accessible through our conscious introspection, the role of neuroscience in this critical approach to moral philosophy is essential. In this context, Peter Singer (2005) stresses the importance of taking a naturalistic approach to ethics. He argues that "recent research in neuroscience gives us new and powerful reasons for taking a critical stance toward common intuitions" (Singer 2005, p.332). The integration of neuroscience and ethics, therefore, can be understood from this critical viewpoint. Levin (2011) and Singer (2005) believe that neuroscience can contribute to critical moral theorizing, i.e., carefully developing moral theories without relying on unjustified intuitions by investigating and isolating them and their hidden influences. It is important to note that in this critical moral theorizing, neuroscience plays not only falsifying roles (filtering out bad intuitions) but also constructive roles of developing or constructing theories of ethics. Greene and his colleagues' (2001, 2004) works on moral neuroscience can be examples of this form of integration. With his brain imaging studies, Greene and his colleagues (Greene et al. 2001, 2004; Greene 2008) criticize intuitive understandings (rational interpretations) of Kantian deontology, in contrast to utilitarianism. Although their arguments do not discuss the whole issues of deontology and are often criticized by others (Liao 2017; Lott 2016), this general approach clearly follows the direction of the constructive integration of neuroscience and ethics in this style of critical moral theorizing through neuroscience. Greene states that "Science can advance ethics by revealing the hidden inner workings of our moral judgments, especially the ones we make intuitively. Once those inner workings are revealed, we may have less confidence in some of our judgments and the ethical theories that are (explicitly or implicitly) based on them" (Greene 2014, pp. 695–696). He believes that neuroscience can go "behind the scenes" of our moral judgments and debunk blindly accepted intuitions and unjustified common conceptions of morality. In so doing, neuroscience can sometimes propose substantial moral claims that can remedy confusing influences of moral intuitions. He states that "I view science as offering a 'behind the scenes' look at human morality. Just as a well-researched biography can, depending on what it reveals, boost or deflate one's esteem for its subject, the scientific investigation of human morality can help us to understand human moral nature, and in so doing change our opinion of it" (Greene 2003, p. 847). It is important to note that Greene (2014, p. 717–725) is not just going behind the scenes to simply criticize faulty or unjustified intuitions. He wants to change some of our intuitive understanding of morality and to defend a particular theory of morality (a form of act consequentialism – moral values of actions and decisions are assessed by their consequences). Greene's approach to empirically informed ethics, therefore, is an inspiring and stimulating example of the constructive integration of neuroscience and ethics through the former's critical analysis of moral intuitions.

The constructive integration of neuroscience and ethics, as I discussed above, is possible and actively pursued in some areas of moral philosophy (Appiah 2010; Churchland 2011; Harris 2010; Kahane 2016; Prinz 2016) and neuroscience (Crockett et al. 2010; Decety and Wheatley 2015; Greene 2014, 2015; Koenigs et al. 2007; Liao 2016; Terbeck et al. 2013; Young et al. 2010).[12] Specifically, there are three possible ways where this integration can be active and promising.[13] To summarize, neuroscience can be integrated with ethics in solving ethical issues in specific contexts, developing normative standards that can apply to a particular group of individuals, and identifying and correcting faulty intuitions. For example, by developing an analysis of how a normative standard of moral agency can include careful consideration of agents' psychological abilities, my discussion of autistic moral agency follows the second way in which neuroscience can be integrated with ethics.

[12] See Berker (2009), Bruni et al. (2014), and Kamm (2009) for critical or more cautious views on the integration of neuroscience and ethics.

[13] There are more gradual and conservative (somewhat skeptical) approaches to the integration of empirical science and normative disciplines. For example, in the context of criminal justice, Morse (2015, p. 74) suggests that one should keep the traditional folk psychological foundation of legal discourse and use neuroscience sporadically if necessary. Morse states that "At present, neuroscience has little to contribute to more just and accurate criminal law policy, doctrine, and individual case adjudication. This was the conclusion reached when I tentatively identified "Brain Overclaim Syndrome" 9 years ago, and it remains true today. In the future, however, as the philosophies of mind and action and neuroscience mutually mature and inform one another, neuroscience will help us understand criminal behavior. Although no radical transformation of criminal justice is likely to occur, neuroscience can inform criminal justice as long as it is relevant to law and translated into the law's folk psychological framework and criteria. The home remedies are working, and please don't wake me until the doctor comes. As Jerry Fodor counseled, "Everything is going to be all right"."

By clearly identifying and distinguishing the neural correlates of cognitive and emotional empathy, one can find a plausible model of empathy that can explain autistic individuals' moral sense and prosocial motivation. I believe that, in this interdisciplinary cooperation, neuroscience and ethics are integrated following the *constructive integration model*. That is, neuroscience does not simply reject implausible moral theories but also fully supports normative theories of autistic moral agency. It plays a *constitutive* role in building a normative standard for autistic agency because normative standards of moral duties and obligations applicable to autistic individuals cannot be meaningfully constructed and explained without considering the peculiar psychological and neurological conditions of autistic individuals.

10.6 Psychological Relevance

According to Kant, a moral duty (as specified by his categorical imperative) should be consistent (logically non-contradictory and conceptually coherent) and universal (being applicable to all rational moral agents) (Kant 1996, G.4.422, 4.434). He states that we should "follow… the practical faculty of reason…where the concept of duty arises from it" (Kant 1996, G.4.412) and "act only in accordance with that maxim through which you can at the same time will that it become a universal law" (Kant 1996, G4.421). These conditions come straight from Kantian moral agency: a rational moral agent has moral duty to follow any rule that satisfies these conditions. It seems that this formal criterion provides an important condition of normative justification of moral principles or values.

> Justificatory Criterion of Normative Standards (1):
>
> *Consistency and Universality*: A normative standard (a standard that guides and regulates human conduct) should be consistent (logically non-contradictory and conceptually coherent) and universal (being applicable to all rational moral agents).

In light of my discussion of neuroscience on autistic moral agency, I add another condition of moral justification. It is the condition of psychological relevance. Normative standards should not only specify ideal values and goals we should pursue and achieve independently of varying contingencies of the world but also provide rules or principles that have good chance of being followed or complied with. One may argue that this criterion of psychological relevance (i.e., a moral principle should be successfully applicable to and effectively practiced by a group of individuals under specific psychological conditions) is not a justificatory criterion of normative values or principles because it depends on psychological contingencies that are not necessarily related to one's rational effort to be moral or ethical. As I discussed above, however, psychological relevance is an important justificatory condition of a normative standard because it refers to a group of individuals with particular cognitive and emotional abilities or disabilities. Considering psychological relevance in justifying a moral standard is not introducing irrational or morally irrelevant contingencies in moral reasoning but bringing ethics to its essential foundation of practicality. A normative standard that cannot be complied with is only an irrational dream or an unrealistic speculation.

Since ideal moral values or normative principles are not just proposed to show their conceptual consistency and universality but to guide and help us to live an ethical life, they should provide psychologically plausible rules and regulations of human conduct. Otherwise, they can become abstract and fancy conceptual exercises that do not lead us anywhere. Normative rules are *prescriptive* rules, rules that are supposed to be practiced, and, for that reason, they should successfully apply to and be followed by a group of agents in their psychological environments. Therefore, I propose the criterion of psychological relevance as a justificatory criterion of acceptable and practical moral values or principles.

> Justificatory Criterion of Normative Standards (2):
>
> *Psychological Relevance*: A normative standard (a standard that guides and regulates human conduct) should provide psychologically plausible rules and values. It should refer to and be followed by a group of individuals with particular psychological abilities.

If a consistent and universal standard of morality does not have psychological relevance, it may not be justified as an acceptable normative standard. Therefore, impractical or irrelevant standards are rejected under this criterion not simply based on an empirical ground but on a normative ground of practical effectiveness.[14] This normative criterion is the key to understanding the constitutive integration of neuroscience in ethics. If a normative standard has psychological relevance in its application and practice, neuroscience plays important roles in the justificatory process of moral values and principles by gauging their psychological effectiveness.

Using empirical data in normative justification is particularly important when one develops a moral standard that relates to and is practiced by a group of individuals. When moral values or normative standards are developed and used in a particular environment, they have to be not only logically consistent and conceptually coherent but also psychologically plausible and practically effective so that they can be followed by individuals with particular moral and cognitive abilities. To be psychologically plausible and practically effective, constructive use of empirical information is necessary. Therefore, neuroscience, as an empirical science of the mind and the brain, is essentially important in developing a psychologically realistic and neurologically plausible standard of morality. Otherwise, a normative standard, even though it is conceptually consistent, does not really work for moral agents. Consider a standard of moral agency such as the fully rational deliberation or cognitive empathy discussed in many theories of moral philosophy. One cannot, however, maintain a *fully* and *exclusively* rational stance in one's moral cognition because of

[14] A similar condition is proposed by Flanagan (1991). My condition of psychological relevance is fully compatible with his principle of minimal psychological realism. Flanagan (1991, p. 32) states that "Make sure when constructing a moral theory or projecting a moral ideal that the character, decision processing, and behavior prescribed are possible, or are perceived to be possible, for creatures like us."

its cognitive-emotional integration in the brain (De Oliveira-Souza et al. 2015, pp. 188–189).[15] In addition, cognitive empathy, as I discussed in previous sections, is not necessary for empathic moral motivation in autistic moral agency. That is, if a theory of moral agency is not psychologically realistic, it cannot function as a practical guideline or be justified as a reasonable and appropriate moral standard.

10.7 Conclusion

In this chapter, I discussed how neuroscience, as an empirical science of the brain, can be integrated into normative discussions of ethics. I explored three models (mutual independence, limited collaboration, and constructive integration) of interdisciplinary interaction between neuroscience and ethics and specified three possible ways neuroscience can be integrated into ethics. I pursued the possibility of the constructive integration of neuroscience and ethics where neuroscience can contribute to the development of normative standards that can refer to a group of individuals under particular psychological conditions. By identifying different types of empathy (cognitive and emotional empathy) and distinguishing their moral psychological properties, I argued that neuroscience can play constitutive roles in developing a normative standard of autistic moral agency. It is observed, in many studies of neuroscience and psychology, that autistic individuals have impaired cognitive empathy but intact emotional empathy that motivates prosocial behaviors. Moral agency of autistic individuals, therefore, can be explained by their intact emotional empathy that facilitates their other-caring and helping behaviors. I also argued that, in developing and justifying a normative standard, its psychological relevance should be considered. Since any normative standard relates to a group of individuals, consideration of that group's cognitive and emotional abilities is critically important. In this regard, integration of neuroscience and ethics can be understood as the theoretical effort to bring neuroscience to the discussion of normative rules and standards that can refer to and be practiced by particular groups of individuals.

There are two normative implications one can draw from the empirical studies of neuroscience discussed in this chapter. First, it is important to consider diverse notions of moral agency (i.e., a normative standard that specifies moral ability and moral responsibility of morally capable individuals). Moral philosophy tends to focus on a small number of moral abilities such as rational deliberation or empathy at the center of moral agency. However, neuroscience shows that there are different types of moral abilities or disabilities one should consider before one praises or blames another's action. For example, autistic moral agency can be explained by emotional empathy without cognitive empathy. The distinction and separation of the two types of empathy is not possible without neuroimaging studies of the brain.

[15] De Oliveira-Souza et al. (2015, p. 189) state that "the fundamental components of morality are neither monolithic nor static but are modified by cognitive-emotional representations and contextual elements as well as by the structure of personality...".

That is, empirical studies of neuroscience are essential in developing, explaining, and utilizing the diverse forms of moral agency in the moral discourse of action and responsibility. Second, moral agency can include partial abilities of the mind. In moral philosophy, a normative standard is typically understood as a standard with fully realized psychological abilities such as rational deliberation or empathy. However, it is possible to think of a moral agent with partially realized abilities. In this chapter, autistic moral agency is explained through emotional empathy without cognitive empathy. The full empathic ability is not present in this form of moral agency. That is, moral agency does not require that the core psychological ability of agency should be fully functional. One can consider partial ability, such as emotional empathy without cognitive empathy, as the core psychological ability of moral agency. All the important points discussed here stimulate an inclusive understanding of the normative standard of moral action and responsibility. They construct a new path for the diverse forms and the partial abilities of moral agency, which neuroscience paves for the future of normative ethics and moral psychology.

References

Aaltola, E. 2014. Affective Empathy as Core Moral Agency: Psychopathy, Autism and Reason Revisited. *Philosophical Explorations* 17 (1): 76–92.

Abend, G. 2013. What the Science of Morality Doesn't Say About Morality. *Philosophy of the Social Sciences* 43 (2): 157–200.

Appiah, K.A. 2010. *Experiments in Ethics*. Cambridge, MA: Harvard University Press.

Armstrong, D.M. 1968. *A Materialist Theory of the Mind*. London: Routledge and Kegan Paul.

Bacon, A.L., D. Fein, R. Morris, L. Waterhouse, and D. Allen. 1998. The Responses of Autistic Children to the Distress of Others. *Journal of Autism and Developmental Disorders* 28 (2): 129–142.

Balconi, M., and Y. Canavesio. 2013. Emotional Contagion and Trait Empathy in Prosocial Behavior in Young People: The Contribution of Autonomic (Facial Feedback) and Balanced Emotional Empathy Scale (BEES) Measures. *Journal of Clinical & Experimental Neuropsychology* 35 (1): 41–48.

Baron-Cohen, S. 1995. *Mindblindness. An Essay on Autism and Theory of Mind*. Cambridge, MA: MIT Press.

———. 2003. *The Essential Difference. Male and Female Brains and the Truth About Autism*. New York: Basic Books.

———. (2009). Autism: The Empathizing–Systemizing (E-S) Theory. *Annals of the New York Academy of Sciences, 1156 (The Year in Cognitive Neuroscience 2009)*, 68–80.

———. 2011. Empathy Deficits in Autism and Psychopathy: Mirror Opposites? In *Navigating the Social World: What Infants, Children, and Other Species Can Teach Us*, ed. M. Banaji and S.A. Gelman, 212–215. New York: Oxford University Press.

Baron-Cohen, S., A. Leslie, and U. Frith. 1985. Does the Autistic Child Have a "Theory of Mind"? *Cognition* 21: 37–46.

Baron-Cohen, S., J. Richler, D. Bisarya, et al. 2003. The Systemising Quotient (SQ): An Investigation of Adults with Asperger Syndrome or High Functioning Autism and Normal Sex Differences. *Philosophical Transactions of the Royal Society* 358: 361–374.

Batson, C.D. 2009. These Things Called Empathy: Eight Related but Distinct Phenomena. In *The Social Neuroscience of Empathy*, ed. J. Decety and W. Ickes, 3–15. Cambridge, MA: MIT Press.

Berker, S. 2009. The Normative Insignificance of Neuroscience. *Philosophy and Public Affairs* 37 (4): 293–329.

Bernhardt, B.C., and T. Singer. 2012. The Neural Basis of Empathy. *Annual Review of Neuroscience* 35: 1–23.

Betti, V., and S.M. Aglioti. 2016. Dynamic Construction of the Neural Networks Underpinning Empathy for Pain. *Neuroscience and Biobehavioral Reviews* 63: 191–206.

Blair, R.J.R. 1996. Brief Report: Morality in the Autistic Child. *Journal of Autism and Developmental Disorder* 26: 571–579.

———. 2005. Responding to the Emotions of Others: Dissociating Forms of Empathy Through the Study of Typical and Psychiatric Populations. *Consciousness and Cognition* 14: 698–718.

Bruni, T., M. Mameli, and R.A. Rini. 2014. The Science of Morality and Its Normative Implications. *Neuroethics* 7 (2): 159–172.

Churchland, P.S. 2011. *Braintrust: What Neuroscience Tells Us About Morality*. Princeton, NJ: Princeton University Press.

Crockett, M.J., L. Clark, M. Hauser, and T.W. Robbins. 2010. Serotonin Selectively Influences Moral Judgment and Behavior Through Effects on Harm Aversion. *Proceedings of the National Academy of Sciences of the United States of America* 107 (40): 17433–17438.

Cuff, B., S.J. Brown, L. Taylor, and D. Howat. 2016. Empathy: A Review of the Concept. *Emotion Review* 8 (2): 144–153.

De Oliveira-Souza, R., R. Zahn, and J. Moll. 2015. Neural Correlates of Human Morality. In *The Moral Brain, a Multidisciplinary Perspective*, ed. J. Decety and T. Wheatley, 183–195. Cambridge, MA: MIT press.

Decety, J. 2011. Dissecting the Neural Mechanisms Mediating Empathy. *Emotion Review* 3 (1): 92–108.

Decety, J., and J.M. Cowell. 2014. The Complex Relation Between Morality and Empathy. *Trends in Cognitive Sciences* 18 (7): 337–339.

———. 2015. The Equivocal Relationship Between Morality and Empathy. In *The Moral Brain, a Multidisciplinary Perspective*, ed. J. Decety and T. Wheatley, 279–302. Cambridge, MA: MIT press.

Decety, J., and P.L. Jackson. 2004. The Functional Architecture of Human Empathy. *Behavioral and Cognitive Neuroscience Reviews* 3: 71–100.

Decety, J., and C. Lamm. 2006. Human Empathy Through the Lens of Social Neuroscience. *Scientific World Journal* 6: 1146–1163.

Decety, J., and M. Meyer. 2008. From Emotion Resonance to Empathic Understanding: A Social Developmental Neuroscience Account. *Development and Psychopathology* 20: 1053–1080.

Decety, J., and T. Wheatley. 2015. *The Moral Brain: A Multidisciplinary Perspective*. Cambridge, MA: MIT Press.

Decety, J., K.J. Michalska, and C.D. Kinzler. 2012. The Contribution of Emotion and Cognition to Moral Sensitivity: A Neurodevelopmental Study. *Cerebral Cortex* 22 (1): 209–220.

Decety, J., K.L. Lewis, and J.M. Cowell. 2015. Specific Electrophysiological Components Disentangle Affective Sharing and Empathic Concern in Psychopathy. *Journal of Neurophysiology* 114 (1): 493–504.

Dodell-Feder, D., J. Koster-Hale, M. Bedny, and R. Saxe. 2011. fMRI Iitem Analysis in a Theory of Mind Task. *NeuroImage* 55 (2): 705–712.

Dretske, F. 1981. *Knowledge and the Flow of Information*. Cambridge MA: MIT Press.

Dziobeck, I., K. Rogers, S. Fleck, M. Bahnemann, H. Heekeren, O. Wolf, and A. Convit. 2008. Dissociation of Cognitive and Emotional Empathy in Adults with Asperger Syndrome Using the Multifaceted Empathy Test (MET). *Journal of Autism and Developmental Disorders* 38: 464–473.

Eisenberg, N., and P.A. Miller. 1987. The Relation of Empathy to Prosocial and Related Behaviors. *Psychological Bulletin* 101 (1): 91–119.

Engen, H.G., and T. Singer. 2013. Empathy Circuits. *Current Opinion in Neurobiology* 23 (2): 275–282.

Fan, Y., C. Chen, S. Chen, J. Decety, and Y. Cheng. 2014. Empathic Arousal and Social Understanding in Individuals with Autism: Evidence from fMRI and ERP Measurements. *Social Cognitive and Affective Neuroscience* 9 (8): 1203–1213.

Felsen, G., and P.B. Reiner. 2015. What Can Neuroscience Contribute to the Debate Over Nudging? *Review of Philosophy and Psychology* 6 (3): 469–479.

Flanagan, O. 1991. *Varieties of Moral Personality: Ethics and Psychological Realism*. Cambridge, MA: Harvard University Press.

———. 2017. *The Geography of Morals: Varieties of Moral Possibility*. New York: Oxford University Press.

Flanagan, O., A. Ancell, S. Martin, and G. Steenbergen. 2014. Empiricism and Normative Ethics: What Do the Biology and the Psychology of Morality Have to Do with Ethics? *Behaviour* 151: 209–228.

Gert, B. 2012. Neuroscience and Morality. *Hastings Center Report* 42 (3): 22–28.

Goldman, A. 1967. A Causal Theory of Knowing. *The Journal of Philosophy* 64: 357–372.

———. 1979. What Is Justified Belief? In *Justification and Knowledge: New Studies in Epistemology*, ed. G. Pappas, 1–23. Dordrecht: Reidel.

———. 1989. Interpretation Psychologized. *Mind and Language* 4: 161–185.

———. 1992. *Liaisons: Philosophy Meets the Cognitive and Social Sciences*. Cambridge, MA: MIT Press.

———. 2012. *Reliabilism and Contemporary Epistemology: Essays*. New York: Oxford University Press.

Grandin, T. 1996. *Thinking in Pictures: Other Reports from My Life with Autism*. New York: Vintage Books.

Greene, J. 2003. Opinion: From Neural 'Is' to Moral 'Ought': What Are the Moral Implications of Neuroscientific Moral Psychology? *Nature Reviews Neuroscience* 4 (10): 846–850.

———. 2008. The Secret Joke of Kant's Soul. In *Moral Psychology*, ed. W. Sinnott-Armstrong, vol. 3, 35–79. Cambridge, MA: MIT Press.

———. 2014. Beyond Point-and-Shoot Morality: Why Cognitive (Neuro)Science Matters for Ethics. *Ethics* 124 (4): 695–726.

———. 2015. The Cognitive Neuroscience of Moral Judgment and Decision Making. In *The Moral Brain: A Multidisciplinary Perspective*, ed. J. Decety, T. Wheatley, J. Decety, and T. Wheatley, 197–220. Cambridge, MA: MIT Press.

Greene, J., L. Sommerville, L. Nystrom, J. Darley, and J. Cohen. 2001. An fMRI Investigation of Emotional Engagement in Moral Judgment. *Science* 293: 2105–2108.

Greene, J., L. Nystrom, A. Engell, J. Darley, and J. Cohen. 2004. The Neural Bases of Cognitive Conflict and Control in Moral Judgment. *Neuron* 44: 389–400.

Hadjikhani, N., N.R. Zürcher, r O. Rogie, L. Hippolyte, E. Lemonnier, T. Ruest, et al. 2014. Emotional contagion for pain is intact in autism spectrum disorders. *Translational Psychiatry* 4 (1): e343.

Haidt, J. 2001. The Emotional Dog and Its Rational Tail: A Social Intuitionist Approach to Moral Judgment. *Psychological Review* 108: 814–834.

———. 2012. *The Righteous Mind. Why We Are Divided by Politics and Religion*. New York: Pantheon Books.

Haidt, J., and C. Joseph. 2007. The Moral Mind: How 5 Sets of Innate Moral Intuitions Guide the Development of Many Culture-Specific Virtues, and Perhaps Even Modules. In *The Innate Mind, Vol. 3: Foundations and Future*, ed. P. Carruthers, S. Laurence, and S. Stich, 367–391. New York: Oxford University Press.

Harari, H., S. Shamay-Tsoory, M. Ravid, and Y. Levkovitz. 2010. Double Dissociation Between Cognitive and Affective Empathy in Borderline Personality Disorder. *Psychiatry Research* 175: 277–279.

Harris, S. 2010. *The Moral Landscape: How Science Can Determine Human Values*. New York: Free Press.

Hauser, M. 2006. *Moral Minds: The Nature of Right and Wrong*. New York: Ecco.

Henry, J., P. Bailey, and P. Rendell. 2008. Empathy, Social Functioning and Schizotypy. *Psychiatry Research* 160: 15–22.

Hoffman, M.L. 2000. *Empathy and Moral Development: Implications of Caring and Justice*. New York: Cambridge University Press.

Hume, D. 1739/1896. In *A Treatise of Human Nature*, ed. L.A. Selby-Bigge. Oxford: Oxford University Press.

Jaarsma, P. 2013. Cultivation of Empathy in Individuals with High-Functioning Autism Spectrum Disorder. *Ethics and Education* 8 (3): 290–300.

Jones, A., F. Happe, F. Gilbert, S. Burnett, and E. Viding. 2010. Feeling, Caring, Knowing: Different Types of Empathy Deficit in Boys With Psychopathic Tendencies and Autism Spectrum Disorder. *Journal of Child Psychology and Psychiatry* 51: 1188–1197.

Kahane, G. 2013. The Armchair and the Trolley: An Argument for Experimental Ethics. *Philosophical Studies* 162 (2): 421–445.

———. 2016. Is, Ought, and the Brain. In *Moral Brains: The Neuroscience of Morality*, ed. S.M. Liao and S.M. Liao, 281–311. New York: Oxford University Press.

Kamm, F.M. 2009. Neuroscience and Moral Reasoning: A Note on Recent Research. *Philosophy and Public Affairs* 37 (4): 330–345.

Kanner, L. 1943. Autistic Disturbances of Affective Contact. *Nervous Child* 2: 217–250.

Kanske, P., A. Böckler, and T. Singer. 2017. Models, Mechanisms and Moderators Dissociating Empathy and Theory of Mind. *Current Topics in Behavioral Neurosciences* 30: 193–206.

Kant, I. (1996). *Practical Philosophy*. M. Gregor (Ed. and Trans.). New York: Cambridge University Press.

Kennett, J. 2002. Autism, Empathy and Moral Agency. *The Philosophical Quarterly* 52: 340–357.

Knobe, J. 2003a. Intentional Action and Side Effects in Ordinary Language. *Analysis* 63: 190–193.

———. 2003b. Intentional Action in Folk Psychology: An Experimental Investigation. *Philosophical Psychology* 16: 309–324.

———. 2006. The Concept of Intentional Action: A Case Study in the Uses of Folk Psychology. *Philosophical Studies* 130: 203–231.

Koenigs, M., L. Young, R. Adolphs, D. Tranel, F. Cushman, M. Hauser, and A. Damasio. 2007. Damage to the Prefrontal Cortex Increases Utilitarian Moral Judgments. *Nature* 446 (7138): 908–911.

Kornblith, H. 1985. *Naturalizing Epistemology*. Cambridge, MA: MIT press.

———. 2002. *Knowledge and Its Place in Nature*. New York: Oxford University Press.

———. 2007. Naturalism and Intuitions. *Grazer Philosophische Studien* 74: 27–49.

———. 2014. *A Naturalistic Epistemology: Selected Papers*. New York: Oxford University Press.

Krahn, T., and A. Fenton. 2009. Autism, Empathy and Questions of Moral Agency. *Journal for the Theory of Social Behaviour* 39 (2): 145–166.

Lamm, C., C.D. Batson, and J. Decety. 2007. The Neural Substrate of Human Empathy: Effects of Perspective Taking and Cognitive Appraisal. *Journal of Cognitive Neuroscience* 19: 42–58.

Leslie, A.M., R. Mallon, and J.A. DiCorcia. 2006. Transgressors, Victims, and Cry Babies: Is Basic Moral Judgment Spared in Autism? *Social Neuroscience* 1: 270–283.

Levin, J. 2011. Levy on Neuroscience, Psychology, and Moral Intuitions. *AJOB Neuroscience* 2 (2): 10–11.

Levy, N. 2009. Empirically Informed Moral Theory: A Sketch of the Landscape. *Ethical Theory and Moral Practice* 12 (1): 3–8.

Liao, S.M. 2016. *Moral Brains: The Neuroscience of Morality*. New York: Oxford University Press.

———. 2017. Neuroscience and Ethics: Assessing Greene's Epistemic Debunking Argument Against Deontology. *Experimental Psychology* 64 (2): 82–92.

Lott, M. 2016. Moral Implications from Cognitive (Neuro)Science? No Clear Route. *Ethics* 127 (1): 241–256.

Marcoux, L., P. Michon, S. Lemelin, J.A. Voisin, E. Vachon-Presseau, and P.L. Jackson. 2014. Feeling but Not Caring: Empathic Alteration in Narcissistic Men with High Psychopathic Traits. *Psychiatry Research: Neuroimaging Section* 224 (3): 341–348.

Marsh, A.A., E.C. Finger, K.A. Fowler, C.J. Adalio, I.N. Jurkowitz, J.C. Schechter, et al. 2013. Empathic Responsiveness in Amygdala and Anterior Cingulate Cortex in Youths with Psychopathic Traits. *Journal of Child Psychology and Psychiatry* 54 (8): 900–910.

Maurage, P., D. Grynberg, X. Noël, F. Joassin, P. Philippot, C. Hanak, et al. 2011. Dissociation Between Affective and Cognitive Empathy in Alcoholism: A Specific Deficit for the Emotional Dimension. *Alcoholism: Clinical and Experimental Research* 35 (9): 1662–1668.

Maurage, P., M. Lahaye, D. Grynberg, A. Jeanjean, L. Guettat, C. Verellen-Dumoulin, et al. 2016. Dissociating Emotional and Cognitive Empathy in Pre-Clinical and Clinical Huntington's Disease. *Psychiatry Research* 237: 103–108.

McGeer, V. 2008. Varieties of Moral Agency: Lessons from Autism (and Psychopathy). In *Moral Psychology*, ed. W. Sinnott-Armstrong, vol. 3, 227–296. Cambridge, MA: MIT press.

Mencius. (2006 – 2017). *Chinese Text Project*. http://ctext.org/mengzi. Last accessed 10 July 2017.

Mikhail, J. 2011. *Elements of Moral Cognition: Rawls' Linguistic Analogy and the Cognitive Science of Moral and Legal Judgment*. New York: Cambridge University Press.

Millikan, R. 1984. Naturalist Reflections on Knowledge. *Pacific Philosophical Quarterly* 65 (4): 315–334.

Montag, C., A. Heinz, D. Kunz, and J. Gallinat. 2007. Self-Reported Empathic Abilities in Schizophrenia. *Schizophrenia Research* 92: 85–89.

Moran, J.M., L.L. Young, R. Saxe, L. Su Mei, D. O'Young, P.L. Mavros, and J.D. Gabrieli. 2011. Impaired Theory of Mind for Moral Judgment in High-Functioning Autism. *Proceedings of the National Academy of Sciences of the United States of America* 108 (7): 2688–2692.

Morse, S.J. 2015. Criminal Law and Common Sense: An Essay on the Perils and Promise of Neuroscience. *Marquette Law Review* 99 (1): 39–74.

Murakami, T., T. Nishimura, and S. Sakurai. 2014. Relation Between Cognitive/Emotional Empathy and Prosocial and Aggressive Behaviors in Elementary and Middle School Students. *Japanese Journal of Developmental Psychology* 25 (4): 399–411.

Nichols, S. 2004. *Sentimental Rules: On the Natural Foundations of Moral Judgment*. New York: Oxford University Press.

Nummenmaa, L., J. Hirvonen, R. Parkkola, and J.K. Hietanen. 2008. Is Eemotional Contagion Special? An fMRI Study on Neural Systems for Affective and Cognitive Empathy. *NeuroImage* 43 (3): 571–580.

O'Brien, E., S.H. Konrath, D. Grühn, and A.L. Hagen. 2013. Empathic Concern and Perspective Taking: Linear and Quadratic Effects of Age Across the Adult Life Span. *Journals of Gerontology Series B: Psychological Sciences & Social Sciences* 68 (2): 168–175.

Popper, K. 1963. *Conjectures and Refutations*. London: Routledge and Keagan Paul.

Poustka, L., A. Rehm, B. Rothermel, S. Steiner, T. Banaschewski, and I. Dziobek. 2010. Dissociation of Cognitive and Emotional Empathy in Autism and Conduct Disorders: The MET-J. *European Child & Adolescent Psychiatry* 19: S80–S81.

Prinz, J. 2004. *Gut Reactions: A Perceptual Theory of Emotion*. New York: Oxford University Press.

———. 2007. *The Emotional Construction of Morals*. New York: Oxford University Press.

———. 2016. Sentimentalism and the Moral Brain. In *Moral Brains: The Neuroscience of Morality*, ed. S.M. Liao and S.M. Liao, 45–73. New York: Oxford University Press.

Quine, W.V.O. 1969. *Ontological Relativity and Other Essays*. New York: Columbia University Press.

Rogers, K., I. Dziobeck, J. Hassenstab, O. Wolf, and A. Convit. 2007. Who Cares? Revisiting Empathy in Asperger Syndrome. *Journal of Autism and Developmental Disorders* 37: 709–715.

Rueda, P., P. Fernández-Berrocal, and S. Baron-Cohen. 2015. Dissociation Between Cognitive and Affective Empathy in Youth with Asperger Syndrome. *European Journal of Developmental Psychology* 12 (1): 85–98.

Schirmann, F. 2013. Invoking the Brain in Studying Morality: A Theoretical and Historical Perspective on the Neuroscience of Morality. *Theory and Psychology* 23 (3): 289–304.

Schurz, M., J. Radua, M. Aichhorn, F. Richlan, and J. Perner. 2014. Fractionating Theory of Mind: A Meta-Analysis of Functional Brain Imaging Studies. *Neuroscience and Biobehavioral Review* 42C: 9–34.

Schwenck, C., J. Mergenthaler, K. Keller, J. Zech, S. Salehi, R. Taurines, et al. 2012. Empathy in Children with Autism and Conduct Disorder: Group-Specific Profiles and Developmental Aspects. *Journal of Child Psychology and Psychiatry* 53: 651–659.

Seara-Cardoso, A., E. Viding, R. Lickley, and C. Sebastian. 2015. Neural Responses to Others' Pain Vary with Psychopathic Traits in Healthy Adult Males. *Cognitive, Affective, & Behavioral Neuroscience* 15 (3): 578–588.

Segal, E.A., K.E. Gerdes, C.A. Lietz, M.A. Wagaman, and J.M. Geiger. 2017. *Assessing Empathy*. New York: Columbia University Press.

Senju, A., V. Southgate, S. White, and U. Frith. 2009. Mindblind Eyes: An Absence of Spontaneous Theory of Mind in Asperger Syndrome. *Science* 325 (5942): 883–885.

Shamay-Tsoory, S.G. 2011. The Neural Bases for Empathy. *The Neuroscientist* 17 (1): 18–24.

Shamay-Tsoory, S.G., J. Aharon-Peretz, and D. Perry. 2009. Two Systems for Empathy: A Double Dissociation Between Emotional and Cognitive Empathy in Inferior Frontal Gyrus Versus Ventromedial Prefrontal Lesions. *Brain: A Journal of Neurology* 132 (3): 617–627.

Sinclair, J. 1992. Bridging the Gaps: An Inside-Out View of Autism (or, Do You Know What I Don't Know?). In *High-Functioning Individuals with Autism*, ed. E. Schopler and G.B. Mesibov, 294–302. New York: Plenum.

Singer, P. 2005. Ethics and Intuitions. *The Journal of Ethics* 9: 331–352.

Singer, T., B. Seymour, J. O'Doherty, H. Kaube, R.J. Dolan, and C.D. Frith. 2004. Empathy for Pain Involves the Affective But not Sensory Components of Pain. *Science* 303: 1157–1162.

Sripada, C.S., and S. Stitch. 2006. A Framework for the Psychology of Norms. In *Innateness and the Structure of the Mind*, ed. P. Carruthers, S. Laurence, and S. Stich, vol. II, 280–302. New York: Oxford University Press.

Stern, R. 2004. Does 'Ought' Imply 'Can'? And Did Kant Think It Does? *Utilitas* 16 (1): 42–61.

Stich, S. 1983. *From Folk Psychology to Cognitive Science: The Case Against Belief*. Cambridge, MA: MIT Press.

———. 1990. *The Fragmentation of Reason: Preface to a Pragmatic Theory of Cognitive Evaluation*. Cambridge, MA: MIT Press.

———. 2006. Is Morality an Elegant Machine or a Kludge? *Journal of Cognition and Culture* 6: 181–189.

Stout, N. 2016a. Reasons-Responsiveness and Moral Responsibility: The Case of Autism. *The Journal of Ethics* 20 (4): 401–418.

———. 2016b. Conversation, Responsibility, and Autism Spectrum Disorder. *Philosophical Psychology* 29 (7): 1015–1028.

Sze, J.A., A. Gyurak, M.S. Goodkind, and R.W. Levenson. 2012. Greater Emotional Empathy and Prosocial Behavior in Late Life. *Emotion* 12 (5): 1129–1140.

Terbeck, S., G. Kahane, S. McTavish, J. Savulescu, N. Levy, M. Hewstone, and P. Cowen. 2013. Beta Adrenergic Blockade Reduces Utilitarian Judgments. *Biological Psychology* 92 (2): 323–328.

Yirmiya, N., M. Sigman, C. Kasari, and P. Mundy. 1992. Empathy and Cognition in High Functioning Children with Autism. *Child Development* 63: 150–160.

Young, L., J.A. Camprodon, M. Hauser, A. Pascual-Leone, and R. Saxe. 2010. Disruption of the Right Temporoparietal Junction with Transcranial Magnetic Stimulation Reduces the Role of Beliefs in Moral Judgments. *Proceedings of the National Academy of Sciences* 107 (15): 6753–6758.

Zaki, J., and K. Ochsner. 2012. The Neuroscience of Empathy: Progress, Pitfalls and Promise. *Nature Neuroscience* 15: 675–680.

Zalla, T., L. Barlassina, M. Buon, and M. Leboyer. 2011. Moral Judgment in Adults with Autism Spectrum Disorders. *Cognition* 121 (1): 115–126.

The manufacturer's authorised representative in the EU is Springer Nature Customer Service Centre GmbH, Europaplatz 3, 69115 Heidelberg, Germany. If you have any concerns regarding our products, please contact ProductSafety@springernature.com

Printed and bound by CPI Group (UK) Ltd, Croydon, CR0 4YY
22/09/2026
02221061-0007